Software Defined Vehicles

Software Defined Vehicles

Plato Pathrose

400 Commonwealth Drive
Warrendale, PA 15096-0001 USA
E-mail: CustomerService@sae.org
Phone: 877-606-7323 (inside USA and Canada)
724-776-4970 (outside USA)
Fax: 724-776-0790

Copyright © 2025 SAE International. All rights reserved.

No part of this publication may be reproduced, stored in a retrieval system, transmitted, in any form or by any means, electronic, mechanical, photocopying, recording, or otherwise, or used for text and data mining, AI training, or similar technologies, without the prior written permission of SAE. For permission and licensing requests, contact SAE Permissions, 400 Commonwealth Drive, Warrendale, PA 15096-0001 USA; e-mail: copyright@sae.org; phone: 724-772-4028

Publisher
Sherry Dickinson Nigam

Product Manager
Amanda Zeidan

Production and Manufacturing Associate
Michelle Silberman

Library of Congress Catalog Number 2025941009
http://dx.doi.org/10.4271/9781468609806

Information contained in this work has been obtained by SAE International from sources believed to be reliable. However, neither SAE International nor its authors guarantee the accuracy or completeness of any information published herein and neither SAE International nor its authors shall be responsible for any errors, omissions, or damages arising out of use of this information. This work is published with the understanding that SAE International and its authors are supplying information but are not attempting to render engineering or other professional services. If such services are required, the assistance of an appropriate professional should be sought.

ISBN-Print   978-1-4686-0979-0
ISBN-PDF   978-1-4686-0980-6
ISBN-epub   978-1-4686-0981-3

To purchase bulk quantities, please contact: SAE Customer Service

E-mail: CustomerService@sae.org
Phone: 877-606-7323 (inside USA and Canada)
        724-776-4970 (outside USA)
Fax: 724-776-0790

Visit the SAE International Bookstore at books.sae.org

In Memory of

Pathrose Augustine Fernandez
My Beloved Father and Companion
"A beacon of unconditional love and support, Someone who believed in me."
Papa! I miss you.

This book is dedicated to:
My father, Late **Mr. Pathrose Augustine Fernandez,** *and mother,*
Mrs. **Telma Fernandez,**
for making me a better person.

My brothers **Plasbo & Pinto**, *who were the guiding stars in my career.*
Sonia & Dew.
Ava, Ethan, & Emma.
Xavier M.X, Philo T.V, Binu, Reshma Rose, Ryan, & Nathan.

My friends, teachers, mentors, and colleagues for their love, motivation, and support.

My beloved wife, **Teena,** *and my son,* **Maximus,** *for their love, encouragement, and patience, and for being the strong pillars of support for me at all times.*

Contents

About this Book xi
Foreword I. xiii
Foreword II xvii
Acknowledgments xix
List of Acronyms xxi

Chapter 01 - Introduction

1.1. Transforming Technologies 2
1.2. Software—A Mighty Double-Edged Sword 3
1.3. Changing Consumer Perspectives 5
1.4. The Needs and Delighters 6
1.5. In Search of Delighters 10
1.6. Summary 11
References 12

Chapter 02 - What Are Software Defined Vehicles?

2.1. The Concept of Systems Thinking 14
2.2. An Overview of the Vehicle and Its Components 16
2.3. Software-Intensive Systems (SIS) and Cyber-Physical Systems (CPS) 18
2.4. Defining SDVs with the Concept of SoS 23
2.5. What an SDV Can Do for You? 27
2.6. Summary 28
References 29

Chapter 03 - Software Defined Vehicles: A Customer's Viewpoint

- 3.1. Functions of Software 32
- 3.2. Software Defined Systems for Users 34
- 3.3. Software Updates and Challenges 36
- 3.4. Understanding Over-The-Air (OTA) Functionality 40
- 3.5. That "WOW" Moment inside the Vehicle 43
- 3.6. How Do They Know I Have a Problem? 45
- 3.7. The Unseen Devil in the Darkness 47
- 3.8. Summary 50
- References 51

Chapter 04 - Software Defined Vehicles: A Manufacturer's Viewpoint

- 4.1. Manufacturer's World and Vehicle Lifecycle 54
- 4.2. Why Should Manufacturers Utilize SDV Concepts? 56
- 4.3. What Can Be Done in Production? 59
- 4.4. What Can Be Done for the Service and Maintenance Phase of the Vehicle? 61
- 4.5. Changing Ecosystem of Regulatory Frameworks 65
- 4.6. How Do SDVs Improve Efficiency? 67
- 4.7. SDVs to Save Vehicle Costs? 69
- 4.8. Use Cases of SDCAVs 71
- 4.9. Challenges and Restrictions 74
- 4.10. Summary 76
- References 77

Chapter 05 - Software Defined Connected and Autonomous Vehicles: An Architecture Overview

- 5.1. Vehicle-Level Architecture of an SDCAV 80
- 5.2. Evolution and Applications of E/E Architecture 84
  - 5.2.1. Distributed Architecture 86
  - 5.2.2. Domain-Centralized Architecture 86
  - 5.2.3. Zonal Architecture 86

  5.2.4. Hybrid Architecture 87
  5.2.5. The Need for a Platform Approach in E/E Architecture 89
5.3. Evolution of Software Architecture and Application. 91
  5.3.1. Monolithic Architecture 91
  5.3.2. Modular-Monolithic Architecture 92
  5.3.3. Service-Oriented Architecture (SOA) 94
  5.3.4. Microservices Architecture 95
  5.3.5. Lambda and Kappa Architecture 97
5.4. An Overview of Vehicle Operating System 98
5.5. Open-Source Software in SDVs 102
5.6. Initiatives to Establish a Standard Architecture for SDVs 103
  5.6.1. Eclipse SDV Working Group 104
  5.6.2. SOAFEE (Scalable Open Architecture for Embedded Edge) 104
  5.6.3. Connected Vehicle Systems Alliance (COVESA) 105
  5.6.4. Hardware Abstraction Layer for Software-Defined Vehicles
    (HAL4SDV) 105
5.7. Summary 106
References 107

Chapter 06 – The Concept of Safety and Security

6.1. General Overview of Safety in Vehicles 110
6.2. Challenges for Implementing Safety Framework in SDVs 112
6.3. General Overview of Cybersecurity in SDVs 117
6.4. An Analysis of Interactions in SDCAVs 122
6.5. An Overview of Safety and Security Concepts for SDCAVs 124
6.6. Summary 130
References 131

Chapter 07 – The Trend Toward Shift-Left and Shift-Right

7.1. An Overview of Traditional Product Development and
  Deployment 134
7.2. Digitalization and Changes from the Classical Product
  Development 136
7.3. The Concepts of Shift-Left and Shift-Right 140
7.4. Tools and Techniques for the Shift-Left Approach 145

7.5. Challenges Faced with the Shift-Left Approach in the Automotive Industry 148

7.6. Some of the Shift-Left and Shift-Right Approaches from the Industry 151

7.7. Summary 154

References 155

Chapter 08 - Transformation in Product Development

8.1. The Shift from Hardware-First to Digital-First 158
8.2. Vehicle as a Development Platform and Data Center 161
8.3. Enablers in Product Development and Testing 165
    8.3.1. AI as an Enabler 165
    8.3.2. Cloud-Based Data Platforms as an Enabler 168
    8.3.3. Connectivity as an Enabler Bridge 170
8.4. Evolving Technology Adoption and Changing Skill Requirements . . . 171
8.5. Summary 173
References 174

Chapter 09 - Future Mobility and Transforming Business Models

9.1. Change from Ownership to a Service 179
9.2. Personalization at All Levels 181
9.3. Changing Business Models and Challenges 182
9.4. Data Monetization 185
9.5. Which Business Model Should an OEM Prioritize? 187
9.6. Digitalization and Ecosystem-Driven Innovation 190
9.7. Time to Market—The Ultimate Goal and Challenges 191
9.8. Summary 193
References 194
Conclusion 195
Index 197
About the Author 201

About this Book

Software Defined Vehicles explains the basics and the general understanding one should have in designing and deploying complex software-dependent cyber-physical systems (CPSs) in vehicles. This book is expected to give readers fundamental knowledge about the concepts of software-defined vehicles (SDVs). Building an SDV not only focuses on the technological complexities and the race of being technologically at the forefront, but there is much more beyond the technology that we see. The mobility ecosystem has transformed over the past few years, and there is a drastic change in technologies that are widely adopted in the mobility industry. Many concepts and technologies from information technology and the consumer industries have penetrated deep into the mobility sector these days. It becomes fascinating once we recognize that most of these changes are driven by the advancements in the technological ecosystem around us from other industries and the evolution of software adoption, rather than the dominance of mechanical or hardware components in vehicles, unlike traditional ways.

Regarding new technologies in vehicles, it is understandable that people have become more lenient in adopting and experiencing them, unlike before. This change in mindset has opened up many new business models and brought in changes to classical mobility businesses. This book will help readers gain the required knowledge and understanding about the once buzzword "software-defined vehicles" and how it will influence their daily lives in this changing ecosystem. This book also puts forward various viewpoints on SDVs with the latest technologies and how they become beneficial and influence users' and vehicle manufacturers' lives and businesses. We are in an interesting era where we can experience an incredible transformation of the mobility sector with advanced technologies and digitalization from various other industries finding their way in.

Foreword I

I have always had a love of cars and all things automotive—the visual impact that a designer imbues in a high-performance sports car, the visceral rawness of a V8 engine at idle, the mechanical complexity of hybrid vehicles, and the modern engineering marvel of electrified vehicles, the most fascinating creations of the modern age. I have been blessed with a 30+ year career working with some of the most talented and creative engineers from all over the world. The process of designing and developing everything from sports cars to off-roaders, to work trucks, and now to electric vehicles (EVs) of all shapes and sizes is at times exciting, frustrating, and rewarding—sometimes all on the same working day. These vehicles, aside from being the second largest investment we make in our lives, create emotional attachments and brand loyalties that bridge cultures, generate passionate loyalties, and perhaps most importantly, move the world safely from one place to another. They are the mechanism by which the post-war world realized the technological and digital revolution and defined the world we live in today.

When Plato Pathrose told me he was going to author this book on SDVs, I was immediately encouraged that the reader would be in very safe hands. Plato has an enviable ability to demystify complex concepts and present them in a way that is easy to consume and analyze. In engineering, we often use storytelling to understand our customers' met and unmet needs. We use this approach to live a day in our customers' lives, understand the frictions they feel, and anticipate their needs. Plato is a highly skilled storyteller with the ability to take many different perspectives and explain the impacts in simple and straightforward terms.

I have had the pleasure of knowing Plato for many years, and I consider him a thought leader in the field of advanced driver assistance systems (ADAS) and automated driving. Since 2006, Plato has worked with a diverse array of professionals in the mobility industry, including vehicle manufacturers, system suppliers, software providers, and engineering service firms. Plato is a lifelong learner, and he practices his professional knowledge by working on real-world problems.

Plato's contributions to the automotive industry are substantial. He has authored several influential books, including *ADAS and Automated Driving - Systems Engineering* and *ADAS and Automated Driving: A Practical Approach to Verification and Validation*. These works have become essential tools for anyone working in the ADAS and automated driving domains, providing valuable insights and practical approaches to system engineering and validation. In addition to his written contributions, Plato has also led cross-industry collaboration groups and set standards for future mobility solutions. His work continues to shape the direction of the automotive industry, driving innovation and ensuring that the transition to SDVs is both safe and effective.

Throughout Plato's enviable career, his storytelling ability has led to him working on some of the car industry's most challenging and complex problems. In this book, he breaks down the intricacies of SDVs, what it means as an engineer, a customer, and a manufacturer. This book zooms into the impact systems thinking has, how modern software has led to breakthroughs in design, the role artificial intelligence and data will play, and why cybersecurity needs to be considered at every step of the design process. This book looks at both physical and functional electrical architectures, describing the design, integration, and software complexity challenges that engineers and software developers face as they build an SDV from concept to reality. The importance of over-the-air updates is discussed in depth, explaining why this capability is now so relevant to us as customers.

The book also explains the role of the vehicle as part of a bigger ecosystem. We will gain insights into how modern computing technologies, including model-based design, cloud computing, and digital twins, can accelerate development timelines and address the exponential increase in validating complex software. We will also discover how the vehicle interacts with its digital environment and various new business models that have been created as we start to realize the opportunity of the SDV as part of a bigger integrated transportation system.

From this book, the reader will gain an appreciation for the complexity and nuances that engineers, designers, and developers face every day in creating this most complex of machines. By demystifying some of the industry terms, Plato has provided a compelling read that demonstrates that an SDV represents a transformative leap in the automotive industry, offering unprecedented opportunities for innovation, efficiency, and customer satisfaction. Plato leaves us with a clear view that the future of mobility is bright, dynamic, and full of potential. The journey ahead promises to redefine our relationship with vehicles, driving us toward a more connected, autonomous, and sustainable world.

Stuart Taylor
Managing Partner and Chief Product Officer—Envorso

Foreword II

After his successful books *ADAS and Automated Driving: A Practical Approach to Verification and Validation,* and *ADAS and Automated Driving - Systems Engineering,* Plato Pathrose is now addressing with his third book another "hot topic" that might have an impact on the automotive industry at least as big as the entry of US and Chinese tech companies into this industry segment with their radically new approaches to vehicle design, architecture and manufacturing: The software defined vehicle.

Actually, the move to the SDV is the next logical step. Vehicles become more and more configurable. In the past, customers could configure their new vehicles on the basis of the pricelist that the dealer provided and order different engines, wheels, a high-end stereo system, or assistance systems such as parking support or distance control. At this time, vehicles had distributed E/E architecture, and with every additional electronically controlled system, an additional ECU was added. This led to complex and expensive architectures, which became very difficult to handle and a source of failures. Domain architectures where one ECU is allocated to each major domain, such as Cockpit or ADAS, solved the problem of complexity and led to a rapid increase in the number of vehicle functions and features that were enabled by a steep increase in computation power that these domain controllers provided. Also, with over-the-air update becoming a standard in modern vehicles, cockpit and ADAS features can regularly be updated and upgraded, new functions can be added, and even subscription models become possible for instance for advanced ADAS functions like hands-free driving on motorways or dedicated services such as the automated search and booking of parking spaces.

However, the configurability of these systems is still rather limited and far away from what we are accustomed to today from our smartphones. The reason for this is that software deployed in a vehicle is still very much linked to the hardware it is running on. Even the latest technology electric vehicles designed and manufactured by tech companies are still far away from being the "smartphone on wheels" that visionary people started to propagate ten years ago, when the Robotaxi craze was at its top.

But new E/E architectures are evolving. Domain architectures will make way for zonal architectures with a few powerful ECUs taking care of certain "vehicle zones." This will lead to an increasing separation between hardware and software, with hardware-agnostic software components as a final result, that can be deployed on different hardware platforms of different manufacturers. The vehicle becomes truly "software defined."

In his book which I had the honor to write the foreword for, Plato describes in an easy-to-understand manner the way towards the SDV and what can be expected from it. The book overviews various use cases and their impact on SDVs and discusses how the influence of digitalization and the technology enablers, such as cloud platforms and artificial intelligence, are changing product development processes in automotive. And last but not least, various business models associated with SDVs are presented.

I hope you enjoy reading Plato's third book as much as I did, and you get an idea of the great potential that lies in this new approach to vehicle hardware and software, which will further reshape the automotive industry.

Matthias Schulze
Global Vice President, ADAS Product—Ecarx

Acknowledgments

This book is designed to help readers understand the concept of SDVs and to explore why the automotive industry is undergoing a profound transformation driven by technologies and ideas originating from other sectors. Before implementing any such change, it is crucial to grasp the underlying principles and motivations. This book aims to provide the reader with the foundational knowledge needed to embark on their software-defined journey with confidence and clarity.

I would like to express my heartfelt gratitude to my *parents* and *brothers* for their unwavering support and encouragement. I lost my *father* midway through this journey, who was very much interested in having this work completed. Thanks for being the light in my life and guiding me all these years. You left me with a story half-told for my next book. Thanks to my beloved wife, *Teena*, for being my first reader and most honest critic, whose insights greatly enhanced this work. I thank my son *Maximus* for his energizing follow-ups and support with illustrations.

A heartfelt thank you to my dear friend *Dhanya D. Rajeev* for your invaluable support with reviews and edits. You went far beyond the role of a reviewer, offering thoughtful suggestions that made this work more accessible, especially for non-engineers. Thanks to *Ethan Pinto, Ameya Kiran,* and *Diya Kiran* for the illustrations and help with the corrections.

I extend my heartfelt gratitude to my mentor Dr. Padmesh, whose guidance has shaped my journey. Your unwavering support in every phase of life means more than words can express.

Special thanks to my friends and ex-colleagues from different organizations for sharing their insights and the way the technology adoption is happening across the industry. It helped me a lot in organizing the topics in this book.

I want to thank *Sherry Nigam* and *Amanda Zeidan* for their support in realizing this book, the editors and reviewers who reviewed the manuscript and recommended updates for improvements, and the entire staff of SAE International for their support. Working with you all was a pleasant experience, and the support I received was amazing.

Above all, I thank *God Almighty* for guiding me and helping me improve in difficult times.

List of Acronyms

ADAS - Advanced Driver Assistance Systems

AI - Artificial Intelligence

ASIL - Automotive Safety Integrity Levels

ASPICE - Automotive Software Process Improvement and Capability dEtermination

AUTOSAR - AUTomotive Open System ARchitecture

BEV - Battery Electric Vehicle

BoM - Bill of Materials

CAN - Controller Area Network

COVESA - Connected Vehicle Systems Alliance

CPS - Cyber-Physical Systems

DSSAD - Data Storage System for Autonomous Driving

E/E - Electrical and Electronics

ECUs - Electronic Control Units

EDR - Event Data Recorder

EoL - End of Line

ES - Enabling Systems

EU - European Union

EV - Electric Vehicle

FDD - Function Driven Development

FoD - Feature-on-Demand

FOTA - Firmware Over-the-Air

GenAI - Generative AI

GigE - Gigabit Ethernet

GPOS - General-Purpose Operating System

GSR - General Safety Regulations

HAL4SDV - Hardware Abstraction Layer for Software-Defined Vehicles

HARA - Hazard and Risk Analysis

HMI - Human-Machine Interface

ICE - Internal Combustion Vehicle

IoT - Internet of Things

IPs - Intellectual Properties

ISA - Intelligent Speed Assist

LIN - Local Interconnect Network

LLMs - Large Language Models

MaaS - Mobility as a Service

NLP - Natural Language Processing

OTA - Over-the-Air

RoI - Return on Investments

RTOS - Real-Time Operating System

SDCAVs - Software-Defined Connected and Autonomous Vehicles

SDS - Software-Defined System

SDV - Software-Defined Vehicle

SIGs - Special Interest Groups

SIS - Software-Intensive Systems

SOA - Service-Oriented Architecture

SOAFEE SIG - Scalable Open Architecture for Embedded Edge Special Interest Group

SoI - System of Interest

SoS - System of Systems

SOTA - Software Over-the-Air

SOTIF - Safety of the Intended Functionality

TDD - Test Driven Development

UNECE - United Nations Economic Commission for Europe

USB - Universal Serial Bus

UX - User Experience

Vehicle OS - Vehicle Operating System

VoC - Voice of the Customer

Chapter 01

Introduction

The evolution of technology has influenced various areas of our day-to-day lives. Mobility is a critical area that is becoming more visible these days. We see different types of vehicles with different features and functionalities compared to age-old classical cars or bicycles. With an engineering mindset, if you analyze this transformation, you can see that the mobility industry is following the advancements in technologies from other industrial sectors such as consumer electronics, information technology, telecommunication, etc. These changes are not always gradual like human evolution. Technological evolution now brings changes much quicker than before due to the growth in technology and better infrastructure. In this book, we will learn advancements in technology in the domain of software-defined vehicles (SDVs). You do not need to be an engineer to understand what is written in this book, nor do you need to go through complex formulas. The goal is to explain technological advancements and the evolution of SDVs in the mobility industry in simple language easily understandable to a student or novice.

1.1.
Transforming Technologies

Any classical electronic system that goes into a vehicle has three main parts: hardware, software, and mechanical. We can see that recently, there has been a huge push and importance given to software as a path forward for technological evolution. Nowadays, vehicle technologies are similar to those we experience in the consumer industry. Connected televisions, mobile phones, smart home technologies, etc., have all found their way into this changing automotive ecosystem.

Unlike in the olden days, life has become fast, and so has the mindset of people. The living environment has changed for many of us, and life now has things that were once a luxury as its fundamental needs. People look for standards and quality of life with these technologies rather than the basic necessity of moving from one point to another. This is the change we see with the transformation of technologies used in the automotive industry.

It was widely said, "Cars are going to become mobile phones on wheels." Undoubtedly, we are too dependent on our mobile phones these days, mainly because of the benefits these small devices bring to our lives beyond the fundamental purpose of making calls. There were days when mobile phones were only used to make calls, and they have evolved now with high-end cameras, audio, internet, gaming, personal assistants, and streaming videos at lightning speeds. These features are now the selling features for a mobile phone. No one will buy a phone now if it is to be sold with only the functionality of making calls.

Now, let us come to cars. What changes can we see from a car we drove 15 years ago to the ones we see now? We have high-resolution screens, video streaming, soft switches and touchscreens, entertainment units for front and rear passengers, etc. Moreover, one of the key attractions is finding a co-pilot or a companion in your car who will assist you, communicate with you for your queries, and suggest actions that help you during your journey. **Table 1.1** presents a comparison between the technologies from the consumer industry, such as a smartphone, and the vehicles we see today.

Table 1.1 Comparison of a smartphone with today's vehicles.

Smartphones	Vehicles
Allow you to make calls, stream videos, and music	Have an infotainment unit that allows calls to be made via connected devices and to stream audio and video
Have voice assistants	Have personal assistants and voice-activated systems
Can add and remove applications and update them from app stores	Allow installation of apps and features (feature on demand) and also updates
Allow over-the-air updates of firmware, operating system, and apps	Allow over-the-air updates and on-demand upgrades for firmware and applications of various vehicle components
Have various sensors such as cameras, gyros, and GPS integrated to provide better performance and functionalities	Have many integrated sensors, including cameras, GPS, and radars for various functionalities

© Plato Pathrose.

The vehicle/car business has changed, and a car is not sold now as a vehicle that will take you from one point to another; instead, it is sold for the features that keep your interests, comfort, and safety during your journey. Sales of cars focus more on options that provide you with a better quality of life during the journey than on the medium that will make that journey happen. This is the power of technology and the depth to which it has influenced your lives. Software is one of the key components in bringing this technology to your devices, vehicles, and the comfort of your house. How these electronic devices can enhance your quality of life with additional comfort, safety, and security is all driven by software. We will also learn how this software changes how we pursue the products and systems that are familiar to us and how it will change our vehicles now and in the future.

1.2.
Software—A Mighty Double-Edged Sword

We have seen how interested we are in transforming our vehicles into consumer tech, like mobile phones. This is possible because of a critical component, software. The definition of software is well known to all from our school days as a set of instructions or programs that we use to operate computers to perform specific tasks. This definition is more important these days as we have transformed ourselves to use and depend on these instructions and programs to perform tasks.

When software becomes the foundation and we become highly dependent on it to perform tasks, it becomes one of the key selling elements for any product in which computers execute specific tasks. As we bring in more and more computational power to our cars, the importance of software also increases in getting tasks executed utilizing the computational power. With the introduction of artificial intelligence (AI) components, the flexibility and performance one could achieve in executing such tasks have increased manifold.

In the new mobility ecosystem, with all the benefits that software offers, such as various functionalities, performance, and flexibility, there is a high risk that it could also have an adverse effect if not taken care of. We see software is in the driver's seat in this transformation, bringing new technologies, connectivity, and enablers of digitalization to the latest vehicles. We believe all will work fine, and we will move ahead in this technological realm with software. That is what our optimistic mind says when we do not deeply understand the technology behind it. The software that provides us with all this flexibility and advancement comes with the risk of taking away our safety and security if not built or deployed with care. We must foresee this when we have all these consumer techs being provided in our vehicles.

As the famous proverb says, "With great power comes great responsibility," which is what we must consider when utilizing software for tech advancements. With vehicles not being just a means of transportation but profoundly influencing our quality of life, our safety and security have become important areas that must be considered in our vehicles. With all its benefits, software will be a mighty sword in the hands of developers and vehicle manufacturers that helps in being competitive, keeping the customer engaged, and extending the product's lifetime. However, this double-edged sword can hurt them if there is a failure in the functionalities they provide and if it has glitches that would compromise the customer's safety and privacy through security flaws. In addition, a newly introduced software could affect the performance of other existing functions. Manufacturers need to be sensible and make sure a technological advancement does not bring risks to their customers. At the same time, it needs to provide a better-quality experience with comfort, safety, and entertainment in the vehicles for customers during their journey.

1.3.
Changing Consumer Perspectives

With the changing ecosystem and technological evolution, not only do products and vehicles change, but people's mindsets also significantly transform, which is primarily driven by the living conditions and life expectancy of the new generation. Earlier, people tended to own luxury objects such as cars that belonged to the family, and they were considered assets that gave them pride in society. With the new generation and the mindset change, one of the significant changes nowadays is that people are less affectionate to the concept of ownership. Using an object or an asset is based on the need to serve a particular purpose. This changed the concept of purchasing the product to purchasing only the service. People are not concerned about the product altogether, but they are strongly concerned about how a particular service is being provided and with what quality.

Another change we see in young customers, one of the most prominent customer groups for the automobile industry, is their willingness and interest in experiencing and trying out new technologies. The younger generation is well updated with the latest technologies and complex systems, starting from mobile phones. They are well-versed in automotive technologies and the functionalities of each vehicle, which makes them ask for specific technologies, such as advanced driver assistance systems (ADAS), connectivity, etc., when buying a vehicle.

People started thinking beyond the concept of a car for their travel. Many micro-mobility solutions, such as e-bikes, pods, etc., facilitate travel in urban areas. Hence, owning a car has reached saturation, mainly because of the difficulty of managing it in urban areas, such as parking, traffic congestion, etc. The possibility of multimodal transportation is common among people now, especially when traveling in urban areas where there are chances of high traffic. This is still not an established way of transportation for countries where the infrastructure is still in development, and the public cannot afford the cost of cars.

Finally, the idea of sustainable transportation and eco-friendly transportation is gaining traction nowadays. People are willing to opt for greener transportation or use sustainable methods. This can range from depending on public transport to using electric vehicles (EVs) for their journey, providing various options and vehicles in the market.

These summarize the recent changes in people's mindsets regarding transportation modes and vehicles. In earlier days, owning a vehicle like a car was not an essential part of someone's life. However, in today's world, many have been thinking of vehicles as a mode of transportation, and that can be attained by opting for services that provide them with the flexibility of movement, such as mobility as a service (MaaS) providers.

1.4.
The Needs and Delighters

The heading might look slightly confusing, but here we are discussing the change in human needs. The human mind is influenced by the quick evolution of various technologies, such as information technologies, which facilitate searching and finding what one needs at one's fingertips within a few seconds.

When we investigate the classical product development and customer satisfaction framework, which was defined by Noriaki Kano in the 1980s, referred to as the Kano model, we can have an interpretation of how products are developed these days and how they are sold with various selling points that keep the customer happy and make them gradually attracted toward the product. A classical Kano model is shown in **Figure 1.1**.

Figure 1.1 Kano model.

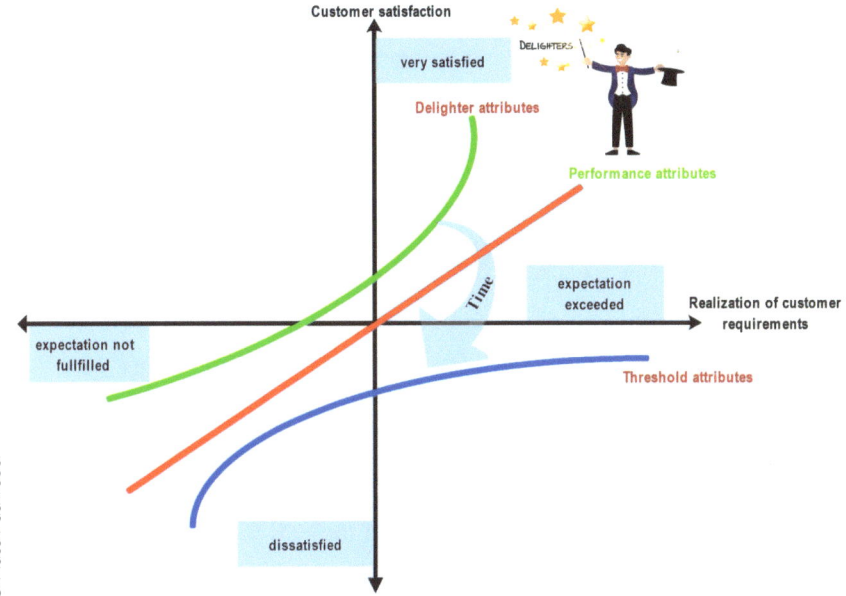

The Kano model will help achieve a thorough understanding of the customer's needs. The model has two dimensions, with the horizontal axis listing the achievements, starting from not so well on the left-hand side to the best or did well on the right-hand side. The vertical axis represents satisfaction, starting from total dissatisfaction on the bottom to total satisfaction as one climbs up the axis toward the top.

Dr. Noriaki Kano has studied and identified three levels of customer expectations, which reflect customer satisfaction with the product's usage [1.1]. These expectations or needs were classified as follows:

1. Expected needs.
2. Normal needs.
3. Exciting needs.

When designing any product or service, the set of use cases that a product or service should address must be defined. Expected needs are the foundational requirements that the product will meet when it is sold in the market. These are usually the entry-level requirements or expectations of such a product or service. They can be defined as certain functionalities, quality needs, attributes, or characteristics.

This set of requirements is critical for the success or failure of the product or service. The customer will not be satisfied by a product or service that has only this set of expected needs or requirements. At the same time, a product or service cannot be successful or even launched without meeting these requirements or addressing these specific expected needs. An example of such requirement types are regulatory requirements. These requirements give identity to the product or address the fundamental use cases, quality, some unspoken expectations when using it, etc. They are also called dissatisfiers for not satisfying the user on their own.

Normal needs are the set of requirements for a product or service that keep it operating in the market. They are a set of requirements around functionalities, quality, and attributes that make the customer satisfied with the product or service. The presence or absence of these requirements in a product or service will have a clear influence on satisfaction or dissatisfaction for the customer. These are usually spoken expectations about a product or service that any product manufacturer lists out when designing a product or service. In the business world, these types of requirements usually integrate with the voice of the customer (VoC) requirements.

Exciting needs, or delighters, are those sets of features and properties that make a product or service a differentiator in the market and a leader. Kano termed these highest-level customer expectations of the product or service's functionalities, qualities, or attributes as "WOW" level needs or delighters.

These specific needs will not affect the launch of a product in the market. On the other hand, they are the key needs or requirements that will position the product or service above its competitors in the market. These delighter requirements will facilitate the likelihood of a customer purchasing the product or adopting the service. They will give the product or service an identity in the market.

These delighter functions will bring back the customer to purchase the product again. Business concepts consider these delighters when launching a new product or service. The customer should not only prefer to buy the product but also return to it or recommend others to buy it to experience those "WOW" features.

Over time, these needs will classically transform to move down the chart, as shown in **Figure 1.2**. Exciting needs or delighters will become normal needs, and normal needs will become expected needs in any product or service. You can see many technological transformations that followed this path when you look around.

Figure 1.2 Transformation of the needs.

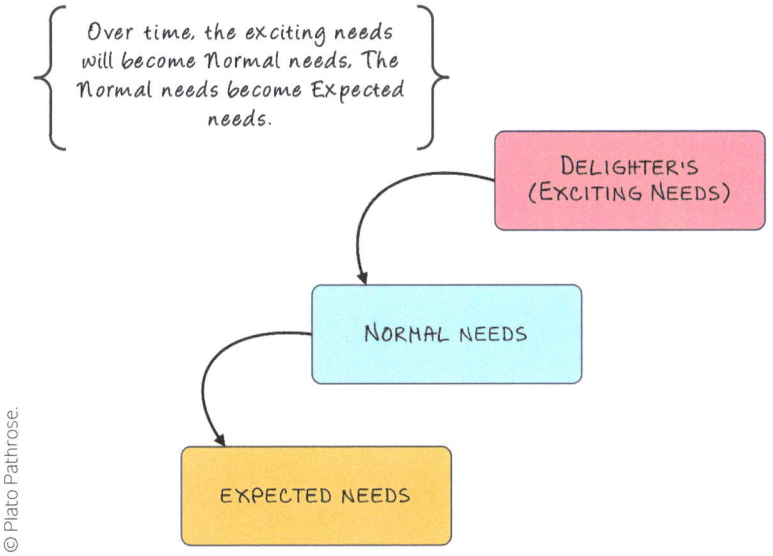

One of the most impressive products is mobile phones with touchscreens. A multi-touchscreen mobile phone became popular in the market with Apple's launch of the iPod and iPhone 1 in 2007, the first phone with a multi-touchscreen. From then on, all mobile phone manufacturers adopted it. We cannot think of a mobile phone without a touchscreen these days. The delighter or the exciting need for mobile phones from 2007 has transformed over the years into an expected need for the product.

1.5.
In Search of Delighters

In today's market, which is deeply driven by various technologies, the delighters or "WOW" factors in any product or service are significantly important. Any organization that gets ahead with technological advancements will attract customers. Innovation is the key here, which drives/motivates customers to buy their products and to adopt their services.

With all the must-have requirements and plenty of good-to-have requirements, the best delighters drive a product and an organization to stay ahead of its competition. This is what we see across various industries. Market leaders have looked at different possibilities to bring such delighters to their customers. It is human psychology that people love changes and "WOWs" more than having the same thing over time. Humans are always interested in and attracted to these delighters. The industry recognized this and found ways to continuously bring delighters to its customers.

In modern cyber-physical systems (CPSs) and devices, software drives most of the "WOWs." This makes them less expensive and easier for customers to manage and experience. Identifying these delighters and continuously making them available to customers keeps the product interesting to them, keeps it competitive in the market, and extends its lifetime or usability time for its customers. This keeps the industry moving forward with the search to continuously identify those "WOW" factors and make them available to its customers in no time.

Understanding and knowing how various delighters or "WOW" factors are identified and made available to users is crucial. The building block for such a business model is the software that makes it possible, without the customer requiring a product change. Due to the transformation of the business and their focus on being market leaders with innovation and technology, many organizations have come up with the strategy of deploying eye-catching features, improved qualities, and enhanced experiences for their products and services throughout their lifecycle through software changes.

With such strategies and business models, organizations have a massive boost in research and software development. How quickly one can identify delighters, build them, and deploy them is the game these days in the technology industry. This creates an innovative ecosystem where businesses search for delighters for their products and services, employing different technologies and building them to make them available to customers in minimal time. This culture impacts the whole product development and deployment lifecycle. The products are being developed in quick iterations, and the changes are adopted more quickly than ever before, employing different development methodologies and processes.

Businesses should also learn from history while searching for delighters or "WOW" features. We read about the multi-touchscreen smartphone launched by Apple in 2007. Surprisingly, Apple's iPhone was not the first one to bring touchscreen phones to market. It was the IBM Simon, which launched on November 23, 1992, the first smartphone with features like a touchscreen, a calendar, a calculator, etc. The product was launched with too many delighters, which was too early to be accepted in the market [1.2]. When businesses search to identify delighters for their products, they should also determine which delighters can bring value to the customers quickly. Having certain delighters in a product or service for which the market is not ready will negatively impact the product. This is what history teaches us with the launch of the touchscreen smartphone by IBM in 1992, compared to that of Apple Inc. in 2007. The technology that was not accepted by the market in 1992 was accepted and became a hit after 15 years.

1.6. Summary

This chapter provided an overview of the changing technology space, ecosystem, and market. It discussed how software has become necessary in the new technological era and supports various business models. Employing the fundamentals from the Kano model, we discussed how change is driven and the search for delighters in the new technological era. This book intends to discuss how various business models and technological changes impact the automotive industry and its path toward deploying SDVs.

The importance of software in bringing changes to vehicles and services offered influences vehicle users. Users can benefit from such vehicles when they use them. Nevertheless, the concept of SDVs and their benefits remains not only with end users but also with vehicle manufacturers and various component suppliers involved in vehicle development and deployment.

This book will give you a clear understanding of an SDV, how it changes business, and how users and manufacturers pursue it. Surprisingly, software plays the primary role in bringing various "WOW" factors and improvements to vehicles. Let us dive into the story behind SDVs.

References

1.1. Kano, N., Seraku, N., Takahashi, F., and Tsuji, S., "Attractive Quality and Must-Be Quality," *Journal of the Japanese Society for Quality Control (in Japanese)* 14, no. 2 (1984): 39-48.

1.2. Sager, I., "Before IPhone and Android Came Simon, the First Smartphone," *Bloomberg Businessweek*, June 29, 2012, https://www.bloomberg.com/news/articles/2012-06-29/before-iphone-and-android-came-simon-the-first-smartphone.

Chapter 02

What Are Software Defined Vehicles?

In the introduction, we learned how software has changed the business outlook and how products are developed and sold in the market. The value of software in a product has increased dramatically, bringing changes and adaptations to it at any point in its lifecycle. In this chapter, we will take a deep look at the importance of software and how it influences the automotive industry. We have been hearing the buzzword "software-defined vehicles," or SDVs, for some time. What are they? Are they real, or are they just another marketing term?

SDVs exist, and they are increasingly being developed as they are expected to be the next big thing in the mobility sector. But how will you define an SDV? This is the trickiest thing we have seen in the industry. This chapter will explain what SDVs mean and what they can do. When we bring bigger terms and buzzwords, it is essential to understand what they mean rather than simply falling for them.

Speaking to many engineers and executives from different automotive organizations, I was surprised that there is no common understanding of an SDV. This was the reason why I started writing this book. This chapter will discuss what an SDV means and how it can be explained from an engineering and business perspective.

2.1.
The Concept of Systems Thinking

Before diving deep into the subject of SDVs, let us see how the fundamental knowledge of systems engineering can help you understand complex systems. SDVs are an advanced approach to systems thinking, where a complex system is seen in depth with its interconnectedness and interdependencies over its entire lifecycle phases rather than just seeing it as an independent component. An SDV is one of those complex systems heavily dependent on software, and this software considerably influences the SDV's functions over its various lifecycle phases.

Defining and detailing a system, as per systems engineering, requires understanding how it will navigate through various lifecycle phases until its retirement. How does a cyber-physical system (CPS) differ from a classical automotive system if it is strongly dependent on software? The thought process on how such software-rich systems navigate through various phases in the systems engineering lifecycle will help us understand the definition of SDVs.

ISO/IEC/IEEE 15288 *Systems and software engineering—System life cycle processes* is a good reference for anyone who would like to learn about a product's various lifecycle phases [2.1, 2.2]. **Figure 2.1** shows the lifecycle phases commonly followed in the automotive industry, covering the re-engineering phase integrated with the standard lifecycle phases in the classical systems and software engineering standard.

Software Defined Vehicles   15

Figure 2.1 An overview of systems engineering lifecycle phases.

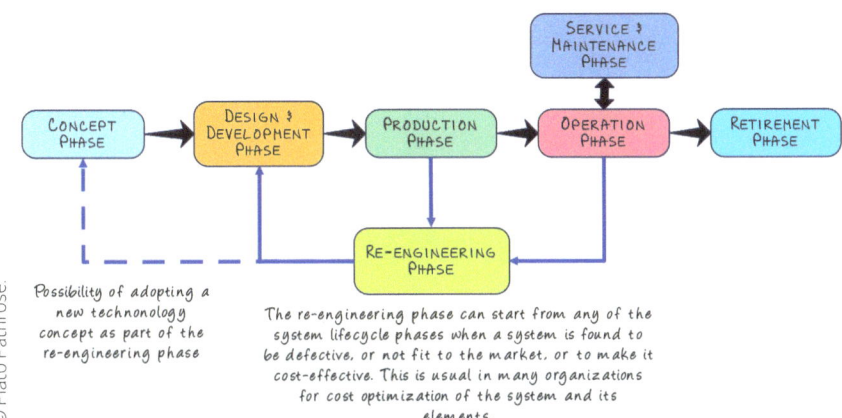

With the power of software in today's software rich CPSs, it becomes possible to manipulate a system's features, quality, and operation in any of its lifecycle phases (**Figure 2.2**). This has become possible because of the in-depth understanding of the interdependence and interconnectedness of various stakeholders and components in a system over its complete lifecycle.

Figure 2.2 Software—the magical power to transform vehicles.

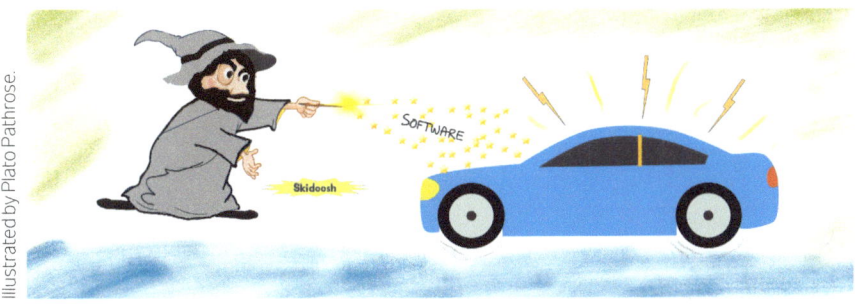

Assume you are a magician who can transform any object with your magic wand. How impressive would that be? These days, software gives vehicle manufacturers this magic power. It acts as the foundation and gives the power to change an existing product in the market.

Systems thinking will provide an overview of the areas impacted and how they will affect various stakeholders associated with the vehicle's complete lifecycle. For a vehicle like a modern car, a systems thinking approach will help you foresee how it will be managed and how various stakeholders and systems inside it will operate during production, when customers use it, when the vehicle is taken to the service center, etc.

If you know how each component operates and how customers will be impacted, you can consider changing any of these around different lifecycle phases. Systems thinking is not just about various lifecycle phases and their impacts; it also involves detailed engineering and business understanding. Engineering knowledge and software integrated into the vehicle and its components will give you the magical power to transform the vehicle in any of these phases of its lifecycle.

We should approach SDVs with a mix of software engineering and automotive engineering, together with systems thinking of the vehicle and its ecosystem. In the following sections, we will see how the magic wand helps change the vehicle and facilitate the rollout of new business models in the mobility industry.

2.2.
An Overview of the Vehicle and Its Components

From the systems engineering point of view, a vehicle can be considered a system, subsystem, or system of systems (SoS) [2.1, 2.2]. It all depends on the boundaries we set. Here, we consider the vehicle as an SoS to better understand SDVs.

Different domains and functionalities associated with the vehicle can be logically partitioned, as shown in **Figure 2.3** [2.3]. Since we are aggressively shifting to EVs and considering highly automated functionalities, the partitions in the below diagram also integrate both these areas associated with next-generation vehicles. From a vehicle's electrical and electronics (E/E) architectural point of view, all these functions are driven by electronic control units (ECUs), domain controllers, or zonal controllers. Software plays a major role in their function and behavior when it behaves as micro- and mini-computers integrated into the vehicle.

Figure 2.3 Vehicle as an SoS.

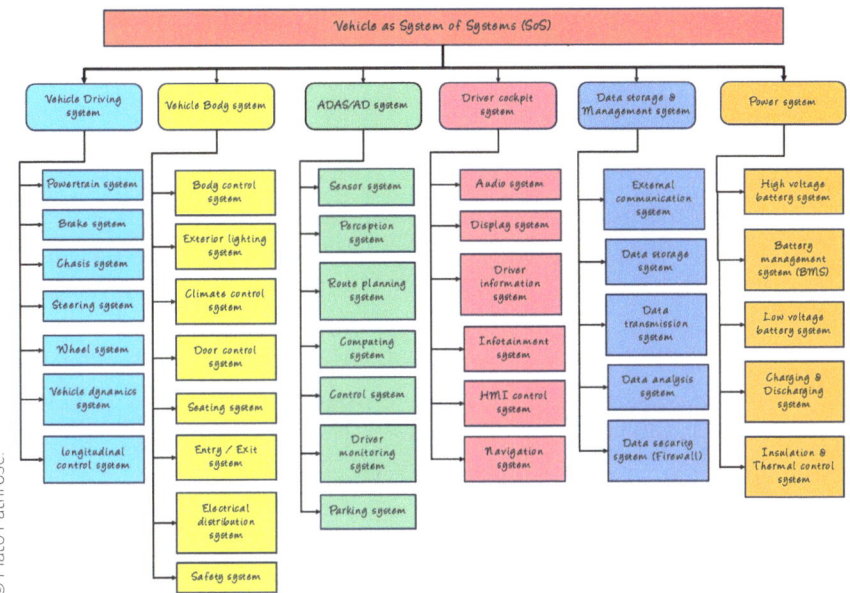

In a vehicle, all these E/E components come from various suppliers for vehicle manufacturers. There could be components developed and integrated by the manufacturers as well. Their functionalities and behavior depend on the software operating on those components. Over the past few decades, we have seen that these electronic components and the software associated with them must be updated or reflashed to fix bugs, adapt the functionalities, and improve functional behavior.

Earlier, we flashed the software in various vehicle components using data cards and CD-ROMs, which later changed to universal serial bus (USB) drives and is now done using wireless methods. What has changed here is the way of software flashing in ECUs and domain controllers in the vehicle compared to the traditional ways. Of late, advancements in the network and telecommunication industry have brought high-speed connectivity with better data bandwidths.

Why have we not called any of these earlier versions SDVs? Software updates are always required and are part of the industry's vehicle manufacturing and deployment process. Even in earlier days, software was an essential part of hardware-rich components in the vehicle. This has changed in recent years with the evolution of electronic components to have more importance for software in today's software-rich CPSs.

If we have so many ECUs and domain controllers in vehicles nowadays driving various functions, which ECU or domain controller must be software-intensive and cyber-physical to consider itself a software-defined system (SDS)? It can be any electronic system integrated as part of the E/E architecture of the vehicle that provides practical functions in the vehicle. These E/E components vary across different vehicles; whether it is an electric vehicle (EV) or a traditional internal combustion engine (ICE) vehicle, the E/E components are there in the vehicle, providing various functions, and software being the critical component, it works together with hardware and mechanical components for the effective operation of the vehicle.

It is always essential to understand a vehicle's E/E architecture when looking at various components in the vehicle and their logical distribution to perform different functions. This will help identify which electronic components influence which functions and recognize the possible scalability of specific domains utilizing various components. Such thinking will give you an early indication of whether the effort in translating certain components in the E/E architecture of the vehicle will have a significant or minimal influence on performing vehicle functions.

2.3.
Software-Intensive Systems (SIS) and Cyber-Physical Systems (CPS)

Software has become an essential part of today's electronic systems. We can see that almost all industries, including the automotive industry, are no different. Traditionally, it was all mechanical and hardware-driven from the technology perspective, with more moving parts. This changed to integrated components on mechanical and hardware segments, totally controlled by software (**Figure 2.4**).

The importance of software has grown from a classical information exchange purpose from different parts of the vehicle to a key component in providing essential functions in the vehicle.

Figure 2.4 Traditional vehicles and modern vehicles.

HARDWARE AND MECHANICAL DRIVEN VEHICLE

MODERN SOFTWARE DRIVEN VEHICLE

With the evolution of technology and electronic components, nowadays we can find two types of systems that are significant to software. In the technology world, it has become very common to use terms such as SIS and CPS. Many have been commonly using them without much clarity on how they are different and what benefits each of them brings to you. We see this getting traction in the automotive industry due to its transformation to a more technology-driven industry of late [2.3, 2.4]. Both are different, and knowing the right meaning will help you understand and use them properly next time.

An SIS is a complex electronic system where software plays a major role in its operations and drives various associated functions. It can be considered a mini-computer on its own, with complex electronic circuits and components associated with it. Since software plays a dominant role, it has a complex architecture and significant processing power to drive the software to bring essential functions and operations [2.4].

Since SISs are complex electronic systems driven by software, they are present in almost all highly complex processing ecosystems, such as cloud computing, data centers, mission-critical systems, etc. They are used across various industries, from information technology, healthcare, and automotive to highly complex defense, space, and avionics. They often focus on providing high-speed and accurate results when information is passed to them. They may be hardwired or plugged into dedicated information exchange and processing ports.

In the automotive industry, SISs are part of specific domain controllers or zonal controllers with one or more processors and operating systems (OSs). They usually operate with complex software architecture and algorithms that work interdependently with other ECUs, domain controllers, or zonal controllers.

A CPS is an advanced version of an SIS. The primary difference here is that CPSs have the important additional functionality of interacting with the external environment. Connectivity plays a major role here for SISs, which are called CPSs. A typical CPS utilizes sensors to understand the external environment. The embedded software operates on various processors and microcontrollers as the primary electronic components in them and has actuators and communication devices to exchange information with the external world. They are widely used for the real-time monitoring and operation of complex functions, with connectivity playing a significant role. Since performance for these systems is very important, they usually integrate AI software components for their operations. This brings the complexity of the software architecture and system to a significant level. It does not mean that AI software components are not used in SISs; however, they do not utilize many AI elements to process environmental data fed via sensors to process in real time. Many of these systems are operational in various industries, such as industrial automation, smart grids, automotive, healthcare systems, and smart cities. An overview of SIS and CPS is shown in **Figure 2.5**.

Figure 2.5 SIS and CPS.

SOFTWARE INTENSIVE SYSTEM

CYBER-PHYSICAL SYSTEM
(A SOFTWARE-INTENSIVE SYSTEM WITH CONNECTIVITY)

One of the commonly seen CPSs in the automotive world is automated driving systems, which provide automation functionality to the vehicle based on the environmental information acquired through various sensors. These systems control the steering and brakes to take the vehicle to a particular destination and establish communication with external components and infrastructure for safe and secure operations.

Now, in SDVs, we have seen how the vehicle's E/E architecture will be when it is decomposed, as shown in **Figure 2.1**. When the vehicle has to be software-defined or has software playing a significant role in its operations and providing functions, these electronic components integrated to form the E/E architecture need to become an SIS or a CPS.

When we have software-intensive CPSs integrated into the E/E architecture, such as an automated driving system, cockpit domain controller, gateway domain controller, etc., they act as the building blocks for the SDV, and these independent systems can also be called SDSs. SDSs are CPSs that act as the foundational or building blocks of SDVs. When not planning to integrate an SDS in the vehicle, we do not expect it to be called an SDV. This is the major difference between traditional vehicles and those we have today in their transformation to SDVs.

SDSs are the building blocks of SDVs. When connectivity and automation are integrated into SDVs, they transform into software-defined connected and autonomous vehicles (SDCAVs) as shown in **Figure 2.6**. SDCAVs have a superset of functionalities integrated into them, where connectivity and highly automated driving functionalities are integrated. Connectivity can be wired or wireless over the complete lifecycle of the vehicle.

Figure 2.6 The evolution of SDCAVs.

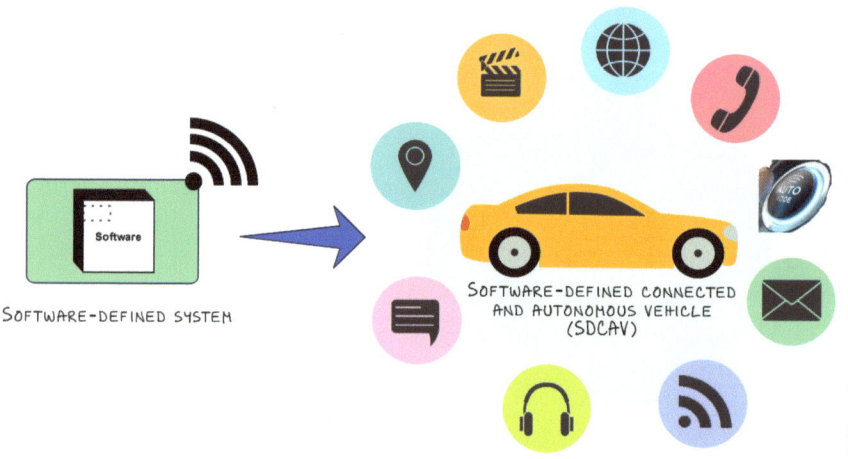

The three key components that define SDCAVs are software, connectivity, and automation. The connectivity of the vehicle to the external environment, such as data platforms, infrastructure, and other vehicles, is driven by specific CPSs, such as modules integrated into the telematics unit or cockpit domain controllers of the vehicle. Depending on the vehicle manufacturer and the capability of various CPSs, different types of connectivity use cases are adopted that bring significant advantages and facilitate the operations for the user and the vehicle manufacturer across the lifecycle phases of the vehicle. How these systems and vehicles transform into a software-defined ecosystem is quite interesting in today's technology world.

2.4.
Defining SDVs with the Concept of SoS

A vehicle comprises multiple systems that are operationally independent but work together to serve the fundamental purpose of mobility and other associated functions. The associated functions bring value to the users of the vehicle by providing comfort, safety, and security while moving from one point to another. Hence, these functions also play a significant role in providing the best experience for the user. Such types of systems are usually known as SoS.

We can consider a vehicle as an SoS with different components or subsystems. At the same time, the whole ecosystem, like a city, can also be considered an SoS. It only differs in the scale at which we define the boundaries of SoS, which incorporates the system of interest (SoI) [2.1, 2.4]. For SDVs, having that system and SoS thinking on a larger scale is always important. One should be able to identify and define various other systems that come into play and facilitate the vehicle's operation during its various lifecycle phases.

When we discuss a vehicle's entire lifecycle, different stakeholders can see how they are involved in various phases. Systems engineering also introduces the concept of enabling systems (ESs).

An ES can be defined as a system or another set of SoS that are not directly involved in the operation of the functional system or subsystems. Instead, it acts as a support system facilitating the functional system's (SoI's) operations in that environment. ESs are not directly involved in the function, but functional systems need them to operate efficiently and effectively. Systems that facilitate the operation and lifecycle activities of the SoI, in this case, systems enabling the lifecycle activities of the SoS, are called ESs. An overview of SoI and ESs is shown in **Figure 2.7**.

Figure 2.7 An overview of the SoI and ES.

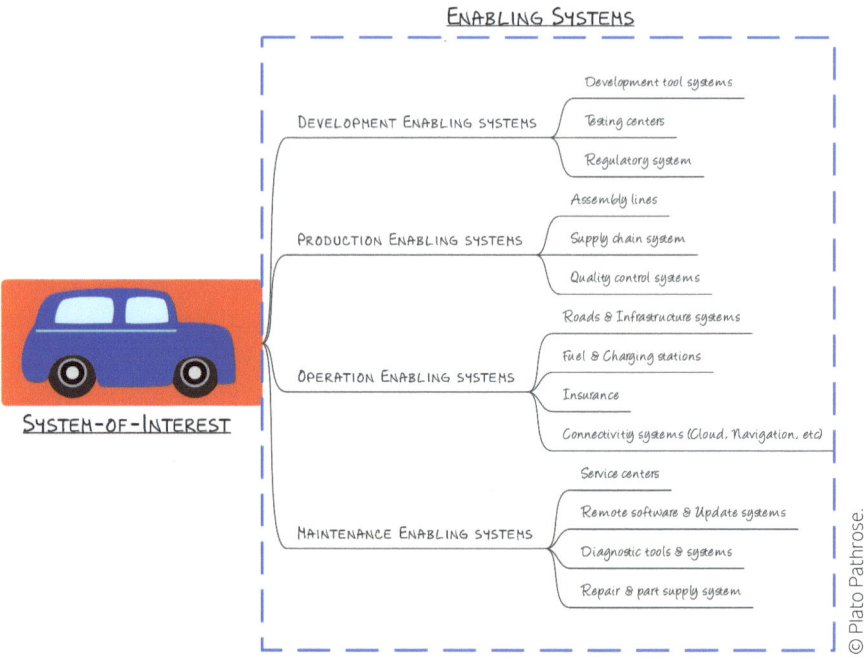

Even though they are present, ESs cannot deliver a particular functional output in the SoS. Instead, they always facilitate the operation of the SoS during its various lifecycle processes. When a vehicle moves from one lifecycle phase to another, these ESs may or may not vary, and they may act as dormant or actively contribute to the efficient operation of the functional system in that specific lifecycle phase.

Identifying the ESs will help recognize which ES is influencing the SoI, the vehicle in our case, during a particular lifecycle phase, and plan the interaction in a better way, employing connectivity and providing additional functions to operate along with the ESs.

For SDVs, software is the driving force that influences the functionality and operation of the system. One would target how software can benefit various stakeholders by utilizing the knowledge of various ESs while defining the use cases and goals for an SDV. Activating certain functions and deactivating specific functions through software configurations or updates at specific lifecycle phases is one way in which this can be achieved. Hence, software updates are one of the key concepts in an SDS and an SDV.

Software updates for each electronic component in a vehicle are not new. Still, the scale at which they are possible now, benefiting from technological advancements and the capabilities of CPSs, helps change the software and adapt it as required on a larger scale. Wireless connectivity and the CPS's high processing capabilities have changed how we did it in the olden days. This is the technological ecosystem that drives SDVs and SDCAVs.

A system or vehicle, with the possibility to have software updates alone, is not the reason to call these components or vehicles as SDSs or SDVs. Another fundamental reason to be called so is how software influences these systems and brings value to its users, manufacturers, or both.

If software can bring value by:

- Improving the performance and functionality of the existing system
- Adding new features and functionalities
- Improving user engagement with the system and the vehicle
- Providing improved safety and security of the system

We can call it an SDV. Software defines how the user engages with the vehicle and provides value additions that improve the quality of experience. **Figure 2.8** summarizes how software can bring value to a vehicle.

Figure 2.8 An overview of software influence in SDVs.

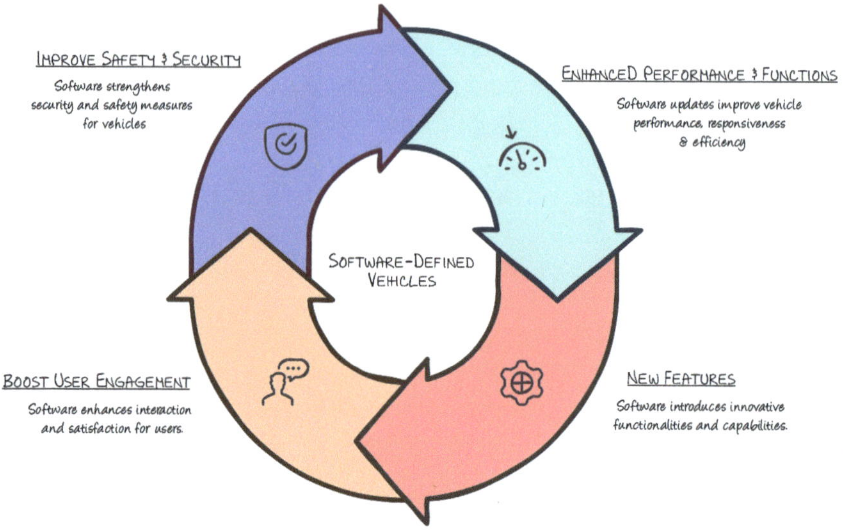

- **IMPROVE SAFETY & SECURITY**: Software strengthens security and safety measures for vehicles
- **ENHANCED PERFORMANCE & FUNCTIONS**: Software updates improve vehicle performance, responsiveness & efficiency
- **NEW FEATURES**: Software introduces innovative functionalities and capabilities.
- **BOOST USER ENGAGEMENT**: Software enhances interaction and satisfaction for users.

© Plato Pathrose.

When looking at a vehicle with many ECUs, domain controllers, or zonal controllers, it becomes tricky for anyone to define which CPS should be capable of software updates and bring value to users. It can be any of those components in the vehicle as long as those systems can bring value to the end user or the manufacturer during various lifecycle phases.

Since we are pushing for greener technology, there is a significant transformation toward adopting EVs and battery technologies. The vehicle need not be an EV to be considered software-defined. As long as it meets the above definition, irrespective of whether it is an EV or an ICE vehicle, it can be considered an SDV.

When discussing SDVs, we always try to think from the end-user perspective. Like the value addition the software can bring to the end user, SDVs also benefit the vehicle manufacturer. By identifying the ESs in the different lifecycle phases, we can see that the vehicle manufacturer operates some of these ESs, such as production systems, data platforms, service stations, etc., directly or indirectly.

SDVs also benefit by having much better interactions with these ESs. The vehicle having specific functionalities and interfaces that are enabled or available only during these specific lifecycle phases will increase the speed of production, performance, and quality of the overall vehicle, which would benefit the vehicle manufacturer in selling the product and gaining a better reputation in the market.

2.5.
What an SDV Can Do for You?

Now that we have learned what an SDV and an SDCAV mean, it would be beneficial to understand what an SDV can do for us. A user can have many benefits, depending on whether the user is the driver or passenger. These benefits are mainly driven by software and vehicle connectivity [2.3, 2.5].

How would you feel when you suddenly have a new safety mechanism activated in your vehicle or a new function that enables you to buy something while sitting inside your vehicle, which you did not have earlier? How would you feel when you can order things directly from your car while you are relaxing in the rear seat, or when you have live updates about your favorite team's match? All these are some of the benefits or advantages that software, along with connectivity, can bring to you while you are in the comfort of your favorite car.

With the evolution of technology, we now have the advantage of having complex CPSs with enormous processing power and storage capacities integrated into our vehicles. The possibility of processing huge amounts of data and sufficient storage can be utilized well using the connectivity of the vehicle [2.5]. Wireless connectivity, such as Wi-Fi and 5G for data and information exchange, will come as a topping on the ice cream, which is already tempting you to taste it.

Nowadays, we usually see SDCAVs with connectivity as their backbone, along with the transition to autonomous driving functions. Even though large-scale applications of autonomous driving functionalities have not happened yet, they have been deployed to serve specific use cases in different industries. Mobility solutions overlap with various other industries such as logistics, industrial automation, mining, etc., where movement of people or material from one point to another can be managed differently from the way it was done earlier, with great advantages.

SDCAVs facilitate mobility in various industries, such as automotive, logistics, marine, and mining. Identifying the right value proposition that software can bring to the vehicle and systems working together with different ESs opens up different business and revenue models. Understanding the engineering and business aspects can help one work on it and develop profitable models over time.

2.6. Summary

This chapter provided an overview and a definition of SDVs. SDV has been a buzzword for some time and has been flexibly adapted to benefit each organization's operational areas. However, it lacks a uniform definition of what it can bring to the automotive industry. In this chapter, the concept of SDVs and SDCAVs has been explained from their building blocks to the scalable SoS concept.

This chapter also presented an overview of how SISs have transformed into CPSs to form the foundational element of an SDV, such as an SDS. In addition, it explained how software has become a necessary element in the new technological era and how it supports various business models. Using the fundamentals from the Kano model that was explained in Chapter 1, how the change was initiated and occurred has been explained in this chapter.

Explaining the systems engineering concept of SoS and ESs helps readers understand how software can bring value to various stakeholders associated with the vehicle during its different lifecycle phases. Software-defined approaches benefit not only vehicle users but also vehicle manufacturers. This has been discussed in the later sections of the chapter.

This chapter clarified some of the confusion one would have in understanding SDVs and SDSs and selecting which of those vehicle systems to be made software-defined. Various business models and possibilities can add value and commercial benefits if proper engineering knowledge and tailored business models are adopted. These have been discussed toward the end of the chapter, opening more possibilities to dive deep into commercialization opportunities of SDVs. The next chapters will help you better understand how SDVs will bring advantages and value addition to users and manufacturers on a large scale.

References

2.1. IEEE SA, "ISO/IEC/IEEE 15288:2015 Systems and Software Engineering—System Life Cycle Processes," 2015.

2.2. Walden, D., Roedler, G., Forsberg, K., Hamelin, R. et al. (Eds), *Systems Engineering Handbook: A Guide for System Life Cycle Processes and Activities*, 4th ed. (Hoboken, NJ: Wiley, 2015).

2.3. Pathrose, P., *ADAS and Automated Driving: Systems Engineering* (Warrendale, PA: SAE International, 2024).

2.4. Lopez-Herrejon, R.E., Martinez, J., Assunção, W.K.G., Ziadi, T. et al., *Handbook of Re-Engineering Software Intensive Systems into Software Product Lines* (Cham: Springer, 2023).

2.5. QNX, "Software-Defined Vehicles," The Ultimate Guide, accessed March 25, 2025, https://blackberry.qnx.com/en/ultimate-guides/software-defined-vehicle.

Chapter 03

Software Defined Vehicles: A Customer's Viewpoint

SDVs and SDSs have been a fancy thing in the market for some time. Many are interested in adopting them and checking whether their vehicles are software-defined. In the previous chapter, we learned about SDSs and SDVs; with all that information, we are moving ahead to see how they are viewed from an end-user point of view.

It is impressive to say that we have an SDV. This chapter will discuss how an SDV will affect an end user's life, whether it will be good, bad, or ugly. SDSs lay the foundation for SDVs. Many vehicle manufacturers started adopting domain controllers as part of their E/E architecture almost a decade ago and have now started migrating to zonal controllers.

With these transitions, they can reduce electronic components, improve connectivity and data exchange, optimize power consumption, and reduce the need for harness and packaging requirements. The most crucial part is that many domains can be consolidated into a single entity in the vehicle with enormous processing power. This brings in many advantages when software influences these vehicle domains over their lifecycle.

This chapter also focuses on understanding how various vehicle users or customers view SDVs. How will an SDV influence their daily lives, and what hidden devil should they be worried about?

3.1. Functions of Software

As we discussed, SDVs are an advanced systems thinking approach, considering the complete lifecycle of the vehicle and software utilized to bring advantages with increasing operational efficiency and delighting customers with amazing features and functions in the vehicle. We should understand the various essential functions the software manages to perform in an SDV to plan and implement such a concept.

Since this chapter specifically focuses on the customer's viewpoint of SDVs, we will mainly discuss the operation phase of the vehicle's lifecycle, where the user or the customer interacts more with the system. However, the operation phase has a close interaction and overlap with the service and maintenance phase of the vehicle. We cover that in detail in the next chapter, where we discuss the vehicle manufacturer's viewpoint of SDVs and how various vehicle manufacturers manage the service and maintenance phase of the vehicle lifecycle.

Software has various critical functions in vehicle operation. Integrating CPSs in the vehicle brings complex electronic systems operating on complex software. Moreover, the connectivity of SDVs with the external world opens many more possibilities for information exchange and software updates without depending on a specific time or location [3.1, 3.2].

For novices, many of these software functions are not concerning if they have various vehicle functions available for their use. However, it is essential to know that complex components work together, utilizing complex software to provide those fantastic functions to the vehicle.

CPSs utilize information from various sensors, analyze it, and act or direct other components to act based on the analysis results, providing various functions to the vehicle. **Figure 3.1** gives an overview of various functions in SDVs driven by software.

Figure 3.1 Various functions of software in the vehicle.

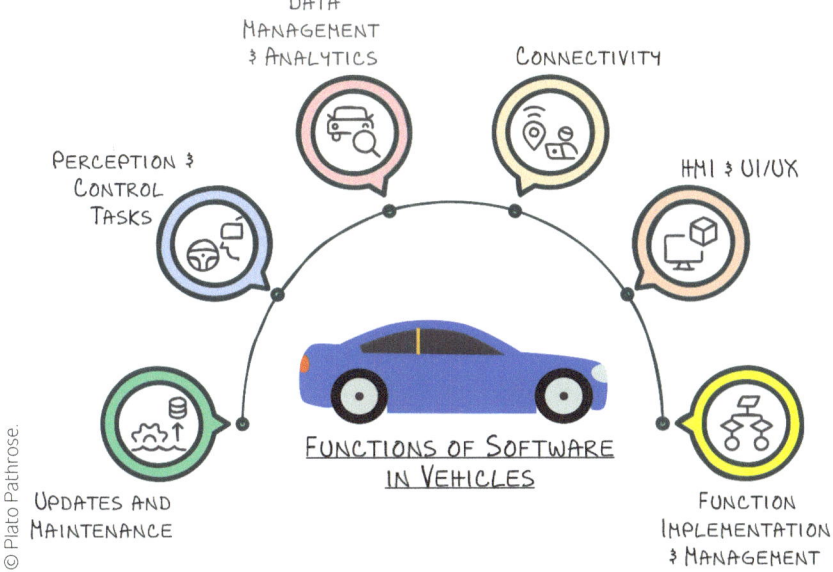

SDVs that integrate CPSs with connectivity have various functions associated with them, as follows:

1. The software takes responsibility for various vehicle functions and manages them for activation, deactivation, prioritization, etc.
2. Human–machine interface (HMI) and user experience (UX) provide users with a better interface to communicate with vehicle controls and have a better experience and engagement while using the vehicle.
3. Perception and control tasks are used to detect various events around the vehicle and decide how the vehicle can be controlled to act on them.

4. The software is involved in data collection, analysis, and sharing, including complete data management from various vehicle components and storing them for diagnostics and analysis.
5. Updates and maintenance of various systems can be carried out through software and communication with the vehicle manufacturer's data platform.

3.2.
Software Defined Systems for Users

We have seen that the fundamental building block of SDCAVs is the CPS called SDS. What would that be like for vehicle users? For those who are not engineers, consider this a system like your smartphone that can connect you with the external world and the digital world, with a lot of data being shared across and used to make you feel comfortable and safe during your journey (**Figure 3.2**).

Figure 3.2 What can smartphones do?

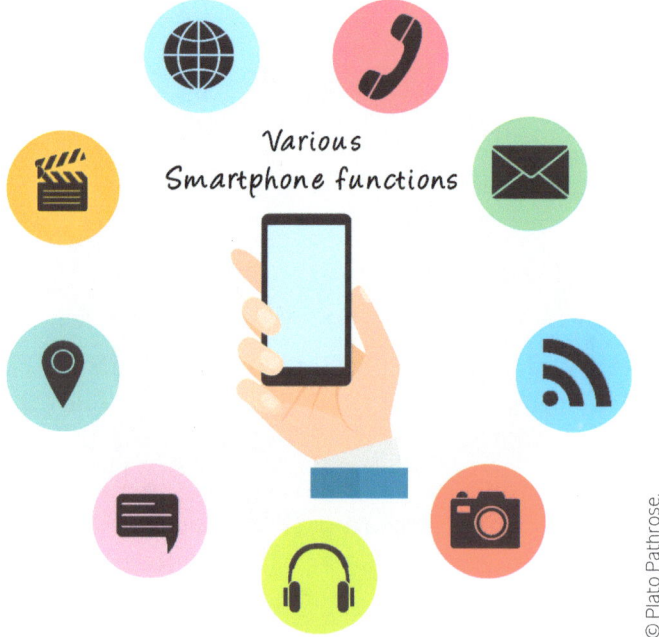

With the advanced concept of SDSs and SDVs, the industry is aiming at executing all the tasks that you have been doing manually or with the support of technology while you are sitting in the comfort of your car. Even though we have seen numerous possibilities for introducing CPSs in various vehicle domains, most vehicle manufacturers tend to introduce software-defined CPSs mainly in two areas with a major impact on their customers. To select which system can be made software-defined and identify its impact on customers, the system should be active and in use for most of its operation time; at the same time, the user should have an active tendency to use its functions while the vehicle is in use. These two characteristics are covered by two main domains of the vehicle: the infotainment domain and the driver assistance domain. It is less likely that anyone will be driving a car without at least playing music or listening to the radio. At the same time, regulations enforce the introduction of certain safety functions in vehicles as part of regulatory frameworks to sell the vehicle in specific regions. This overlaps with displaying information and notifications to the user while driving. This brings in the value of various display mechanisms inside the vehicle, such as dashboard display, instrument cluster, head-up display, etc.

To influence customers, vehicle manufacturers have introduced various additional functionalities associated with these modules in the vehicle, with which customers often interact. This is like providing the interface of a smartphone inside the vehicle, with possibilities to do shopping, pay parking tickets, order food, browse the internet and social media, and make calls. How would it be if customers could install additional applications to access various services from the internet? Many vehicle manufacturers have started providing this function to their users: app stores where users can even install, access, and play online games while sitting inside the vehicle.

How do you feel about the possibility of buying safety features for your vehicle via app stores from your vehicle dashboard? It seems interesting, right? Yes! It is also possible to make payments and enable certain features in the vehicle. All these are driven by SDSs that are closely connected systems managed by complex software with connectivity to the vehicle manufacturer's backend data centers. These SDSs for vehicle customers are next-generation vehicle technology. They give the flexibility to users to perform many tasks while sitting in the comfort of the vehicle.

3.3.
Software Updates and Challenges

One key strategy for SDVs is their open approach to software updates in any CPS integrated into them. The power of high-speed and high-bandwidth connectivity allows the sharing of a huge amount of data with the vehicle and from external infrastructures. This brings flexibility to SDVs as there is no dependency on time and location to provide software updates or configuration changes in the vehicle once triggered, if connectivity is available.

Software updates are the major processes in SDVs for bringing changes to the vehicle software by adding new features, improving the performance and quality of existing functions, and debugging and fixing faults. Software updates are also a way to delight customers while they use the vehicle. Are they only possible with connectivity? Or something else as well?

Nowadays, software updates in vehicles are usually performed using two methods: the direct method and the indirect method. The direct method utilizes a wired or wireless connection between the vehicle and a data source where the software repository is available. It could be a data storage device plugged into the vehicle with software packages. The wireless method has a direct wireless network connection, such as WiFi or 5G, from the vehicle to the vehicle manufacturer's server, and software packages should be downloaded and stored in the vehicle for various vehicle components that must be updated. This is widely called over-the-air (OTA) updates.

The indirect method utilizes a third-party device as an intermediate source to transfer software packages to the vehicle. This includes a direct wired connection to a PC with software packages downloaded or a mobile device such as a smartphone, downloading the software earlier from the vehicle manufacturer's server, and initiating the transfer and update while it is connected wirelessly to the vehicle.

Both vehicle manufacturers and users should be aware of a few things regarding software updates to the vehicle and its components. The update process starts with the vehicle manufacturer integrating the CPSs in the vehicle and understanding how they are expected to operate over their lifecycle and how software updates will be performed. Having high-speed interfaces between these CPSs is the key factor here, as the software size and update time required for these systems depend on these interfaces and how well they are integrated into the vehicle.

Updating navigation maps is one of the most common use cases for almost all vehicles these days. When the scope of software updates goes toward infotainment units and driver assistance systems, updates should be made considering all possible impacts. Moreover, software quality and rollback possibilities are key to managing and addressing the drawbacks while they are being updated. Software update strategy is critical and should be considered from a vehicle manufacturer's viewpoint. Software updates should be planned precisely regarding when, how, and which components should be updated.

Assume the vehicle systems are being updated while the vehicle is in use, and critical features are unavailable. This is a bad strategy; unfortunately, many vehicles in the market follow this bad software update strategy (**Figure 3.3**). This will do more harm than benefit to the user.

Figure 3.3 Software updates while the vehicle is in use.

Software updates through direct and indirect methods are good if executed well. The transformation of vehicles to green energy sources and batteries also affects this software update strategy. We should understand that vehicles do not have an unlimited source of power that can be used for software updates for an infinite amount of time. The vehicle shutdown states, sleep state, and all power cycle states from various power sources in the vehicle should be evaluated and planned while defining the software update strategy.

For a user unfamiliar with software updates or experiencing them for the first time, verifying the software package used for updates is always recommended. If it must be done manually, wrong software packages and incorrect update procedures can corrupt electronic systems and brick them, similar to a smartphone. This could also give cybercriminals a way to access your vehicle and the information stored in it. Moreover, breaking the system in the vehicle would bring additional service and repair expenses that may not be good to spend out of your pocket. This is because software update strategies are not standardized, and each vehicle manufacturer has their own approaches, even though they all have CPSs in their vehicles.

Even with these risks, there are advantages to these OTA software updates. They reduce the dependency on technicians for any software updates and, if initiated from the vehicle manufacturer's side, will be error-free compared to manually updating this in service centers. With many vehicle manufacturers advancing in OTA software updates, it has become more like a daily routine to have software availability and push for updates when there is a need for any software updates to a component in the vehicle.

One key factor driving SDVs is the interdependency of hardware and software in them. This also opens up the scope for the development and deployment of software without depending on hardware each time. At the same time, vehicle manufacturers must ensure that the hardware they put in the vehicle is mature and can manage technology trends and lifecycle for the next 10 to 12 years. Having high-end electronic components in the vehicle will add to the cost of the vehicle. How can this be managed?

Some vehicle manufacturers and their suppliers have devised a solution to control vehicle costs and reduce hardware bill of material (BoM) costs. Operating certain complex software that requires high processing power and memory outside the vehicle in a data platform, such as a cloud platform. It has been a successful solution to execute the function by transferring the sensor data remotely to a data platform and executing most tasks requiring high processing power via vehicle connectivity. This will reduce the need to integrate high-end processors and memory modules into CPSs, thereby reducing the cost of these vehicle components.

Using this approach, certain vehicle manufacturers have deployed some of the low-speed automated driving functions, such as autonomous parking and valet parking, in their vehicles and enabled them through software updates. Rather than processing the sensor data with complex software components within the vehicle, they process them on a remote data platform. Even though this approach adds dependency on high-speed connectivity and availability, it successfully reduces the cost of vehicle components, which usually falls to the end customer as the cost of the vehicle. However, this solution adds a dependency on the availability of a high-speed internet connection in the vehicle. Software updates can also bring these concepts to life if they have been approached with a better vehicle design strategy.

The industry must standardize software updates and their management. The current standards and regulations are more biased toward cybersecurity needs, such as United Nations Economic Commission for Europe (UNECE) regulations R155 and R156 [3.3, 3.4]. However, the workflow and strategies still differ widely across vehicle manufacturers in terms of how they are implemented. In the following chapters, we will discuss the technical dependencies and challenges in software updates in SDVs from an architectural perspective.

3.4.
Understanding Over-The-Air (OTA) Functionality

Over-the-air (OTA) refers to the software update method in CPSs that uses wireless communication. It is also used for configuring devices, exchanging encrypted data, sharing system status and diagnostics information, and more.

SDVs utilize OTA as a key feature for updating and upgrading various CPSs within the vehicle. OTA enables updates for both firmware and application software, offering the flexibility to enhance vehicle performance and features. However, the concept of OTA is not limited to SDVs; it applies to any CPS that supports data access and wireless communication with a data platform. The technology originated in the telecommunications and Internet of Things (IoT) domains. From the SDV perspective, the use cases for OTA are vast. It enables remote software updates, configuration or setting changes, security patch deployment, bug fixes, feature rollouts, and vehicle health diagnostics to support proactive service and maintenance.

In the automotive industry, OTA can be broadly categorized into two functions: firmware over-the-air (FOTA) and software over-the-air (SOTA). FOTA, typically provided free of charge, includes updates to existing software in the vehicle's CPSs, such as firmware updates, software bug fixes, security patch deployments, and configuration adjustments. This improves the reliability and functionality of existing vehicle systems.

SOTA, on the other hand, refers to the OTA function that replaces or modifies software components in the vehicle CPSs. It is commonly used to introduce new functionalities or services, often offered by manufacturers as paid subscription-based features. Many feature-on-demand (FoD) services are delivered via SOTA, allowing users to unlock and activate specific features during the vehicle's operational lifecycle. **Figure 3.4** presents an overview of an OTA architecture used in the automotive industry.

Figure 3.4 An overview of OTA architecture.

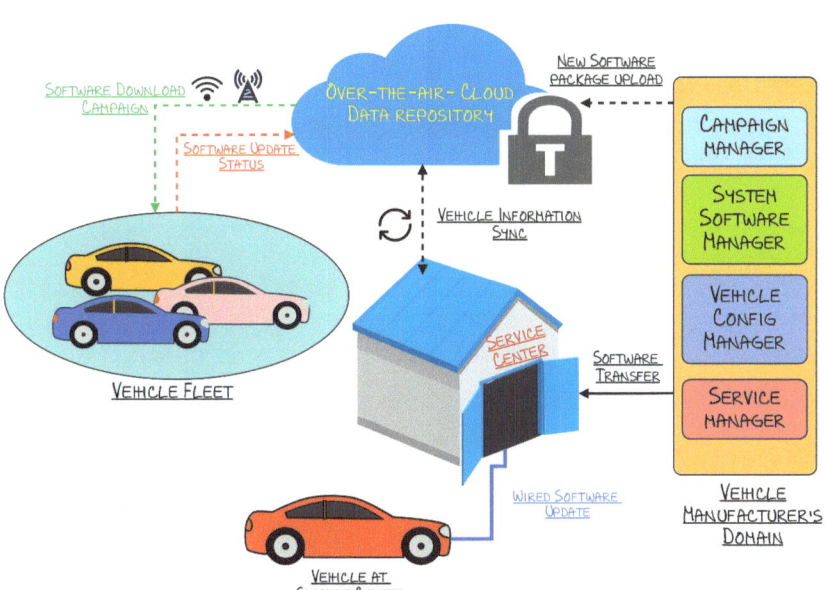

The process begins with identifying the necessary software updates and creating the update package, which may include bug fixes, performance improvements, security or safety patches, and configuration changes. Software can be sourced from suppliers, provided it has passed validation and is ready for deployment. The software is then packaged, typically encrypted, along with all required metadata and settings for installation. This package is uploaded to a server for distribution to vehicles or CPSs.

In SDVs, these packages are often uploaded to a cloud-based data platform or server that manages distribution across multiple vehicles. Updates can be detected either directly by the vehicle through periodic server checks or by a companion app installed on the customer's smartphone. Depending on the configuration, packages may be downloaded automatically or require manual initiation by the user, either within the vehicle system or via a connected smart device.

Downloaded packages undergo authentication and integrity checks to guard against cyberattacks or tampering. After validation, the installation process begins. Post-installation checks are conducted to identify any issues or malfunctions, and rollback mechanisms are provided if problems are detected. Update strategies vary among manufacturers, depending on vehicle type, update size, system connectivity, and update duration.

The OTA architecture in automotive involves several key modules managed by the vehicle manufacturers, such as:

- **Campaign Manager:** Oversees the entire OTA update campaign, whether standard or mandatory updates for specific vehicle systems.
- **System Software Manager:** Handles software packages and part numbers for each electronic component in the vehicle. It manages software changes and updates within the vehicle.
- **Vehicle Configuration Manager:** Manages vehicle-specific configurations across models and variants, crucial for fleet-level updates.
- **Service Manager:** Ensures software updates are available at service centers and dealerships, where they can be flashed into serviced, new or unsold vehicles via wired or wireless means.

OTA provides numerous advantages beyond the automotive industry, finding applications in consumer electronics, IoT devices, and more. It eliminates the need for physical access to devices, allowing for remote software management without manual intervention. OTA offers faster response times for bug fixes and security updates, improved scalability, and reduced maintenance costs. These benefits contribute to a better user experience and position OTA as a critical functionality in SDVs and other CPSs across different industries.

3.5.
That "WOW" Moment inside the Vehicle

We learned that SDVs bring better experiences for their users. In the first chapter, we discussed delighters from the Kano model and the vehicle manufacturer's journey in finding and bringing in "WOW" moments to the vehicle so that they can excite their users and enrich their experience with the vehicle. It is also a way to improve the engagement of vehicle users and be successful in the business by keeping the customer bound to a specific vehicle brand.

What could be these "WOW" moments for a driver or a normal vehicle user? They can be anything from a small birthday wish from the vehicle on your birthday to a fantastic feature enabled in your vehicle through payments or as a gift. Any event or instance becomes a "WOW" moment when you get it when least expected.

Vehicle manufacturers have been continuously striving to incorporate delighter functions inside the vehicle to promote SDVs and keep their customer base solid and expanding. The focus areas of these delighter functions that vehicle manufacturers target can be summarized as follows:

1. Enhanced UX and improved satisfaction.
2. Establishing an emotional connection with the brand and loyalty to it.
3. Enhancement in social and entertainment features in the vehicle.
4. Improvement in safety and security.

5. Functional improvements and optimizations for performance and quality.
6. Social appeal on luxury and standard of living.

To target these areas, the focus should be on vehicle components that come into direct contact with the customer or vehicle functions that a user frequently uses while in the vehicle. Psychologically, targeting things that are closely connected or in contact with people can have more influence on people. Hence, the best area to target would be the cockpit domain of the vehicle, where there are displays, switches, and an interface for communicating with the vehicle by touch, gesture, or voice commands. That is how personal assistants powered by AI, which are common in today's vehicles, became a "WOW" factor when they were launched.

The story continues even today, as vehicle manufacturers target their customers to reach them with exciting features, birthday wishes, and improvements in the way the user engages with the vehicle. Infotainment systems are displays targeted to improve the UX and interaction, bringing mental satisfaction to the user. Many vehicle manufacturers have also started focusing on providing the facility to do specific tasks while the user is sitting inside the vehicle, such as paying parking tickets, shopping, ordering food, etc.

Nowadays, a few vehicle manufacturers have started integrating the possibility for users to purchase, activate, and deactivate certain features in the vehicle directly by making payments for those services. These features extend from subscribing to some music applications to safety-critical autonomous driving functionalities, such as highway pilot or traffic jam pilot, in the vehicle (**Figure 3.5**). With technological advancements, AI components are being used significantly in CPSs, which opens immense opportunities and ways to interact with vehicle users.

Do you not think having a birthday wish on your birthday and getting the autonomous driving functionality activated in your car as a gift would be a "WOW" factor for you?

Figure 3.5 The "WOW" factor from the vehicle.

3.6.
How Do They Know I Have a Problem?

Like the "WOW" factors we discussed before, vehicle manufacturers try to delight customers by utilizing the power of AI in the CPS integrated into the vehicle. We learned that CPSs utilize information from various sensors, analyze it, and act on that information directly or together with other components in the vehicle to provide useful functions. Remote diagnostics and predictive maintenance are functions that come as part of the service and maintenance phase of the vehicle lifecycle, but they overlap with various functions in the vehicle's operation phase.

Predictive maintenance functions in vehicles utilize advanced data analytics and information from various vehicle sensors to analyze and predict a possible failure or a requirement for the maintenance of vehicle components (**Figure 3.6**). They widely use machine learning (ML) algorithms as their foundation to perform data analytics and decision-making.

Predictive maintenance brings many advantages for a vehicle or a fleet of vehicles, including preventing unexpected breakdowns and improving the safety and cost savings for users. It also increases the vehicle's lifespan by identifying and reporting possible issues early and, in some cases, directly initiating and triggering service calls from the vehicle to the service center.

Figure 3.6 Predictive maintenance in the vehicle.

Remote diagnostics is a service and maintenance activity that allows remote diagnosis and analysis of the vehicle status and failures. This saves manual inspection and evaluation, so that taking the vehicle to the service center is not even required to solve certain failures.

Remote diagnostics functions have a higher dependency on connectivity from the vehicle to the vehicle manufacturer's backend data platform or server. Vehicle sensors will capture the status and information from vehicle components and transmit them to the vehicle manufacturer's server, where analysis and diagnostics will identify the issues and propose a configuration change or initiate a new software package release to resolve the issues in the vehicle.

To provide these functions in an SDV, the basic requirement it should possess is connectivity. CPSs with ML algorithms deployed on them can analyze the collected data and perform specific functions. It can be managed up to a level without the need for very high-speed and high-bandwidth connectivity. However, remote diagnostics and immediate fixes of the issues via software updates would be efficient only if connectivity is good enough to manage the traffic of data exchange between the vehicle and the vehicle manufacturer's data platform.

It will be an interesting experience, or even a "WOW" factor when the vehicle knows that a failure is expected to happen soon and can warn the driver. On the other hand, this also increases the complexity of the systems inside the vehicle and the need for complex software elements to operate continuously and communicate with each other and with the vehicle manufacturer's data server.

3.7.
The Unseen Devil in the Darkness

With enormous advantages of SDCAVs for the end user, from better safety, security, and quality functions to those "WOW" factors, SDSs have transformed the automotive industry. However, with all those flashy features and benefits, there are also reasons why we should be concerned about these new technologies and approaches in the automotive industry.

The concern is mainly around the growth of and strong dependency on software in vehicles. Nowadays, vehicles have transformed to be driven by software rather than hardware or mechanical systems as in the olden days. They have become computers or smartphones on wheels with the features and flexibility they provide to the end user.

Issuing software updates to the vehicle is the main method to bring in changes in an SDV; this is a significant flaw in many vehicles. When the vehicle manufacturer depends on an enormous number of suppliers for software, planning and deploying updates on each vehicle over the vehicle's lifespan can become a complex process. Getting into a software-driven mentality also forces vehicle manufacturers to deploy their vehicles with partially tested software, anticipating that a better version of the software can be made available within weeks before the vehicle reaches the customer. In most cases, such anticipations fail, and the customer may not have a stable or good version of the software in the systems. This is dependent on the software release strategies of various vehicle manufacturers. There are certain vehicle manufacturers with really good strategies, where software deployment plans and fixes are preplanned, and even inform the customer about a planned software update, and certain others that take *ad hoc* approaches in deploying software with poor plans and strategies from the market.

We have discussed that certain vehicle manufacturers carry out data processing outside of the vehicle to a remote data platform to save costs. Especially for safety-critical functions, this creates an unwanted dependency on vehicle connectivity. It would seriously affect the availability of those functions when there are glitches in connectivity. At the same time, this brings in dependency on subscribing to the internet for the vehicle's lifetime if they want to use this feature.

The major issue with many vehicle manufacturers is the lack of importance given to the integration testing of various vehicle components when the software undergoes an update. Software packages deployed as updates might not undergo an extended testing period like the production software in the vehicle to reduce engineering expenses. This can bring in major issues, such as the software update affecting other functioning systems or even causing vehicle malfunction (**Figure 3.7**). Further analysis and updates must be planned to fix them, which can take another few days or longer.

Figure 3.7 Some bad things about software in SDVs.

With the difficulties surrounding software updates, which we discussed earlier, there comes another concern that usually alarms users. With the current ecosystem of many emerging vehicle manufacturers, the software update practice happens in their vehicles due to their weak processes around software management. This results in frequent random software updates, planned and unplanned, which may be to fix specific bugs or to improve the performance of existing functions in the vehicle. Frequent software updates to the vehicle make it difficult for the user as they will prevent the vehicle from being used in many cases if the update strategy and software deployment processes are not well planned. With the software package size that must be downloaded and flashed on various vehicle systems, the time taken for those systems to become functional again is affected, which can affect the customer's interest in the product.

However, dependency on connectivity and storage of private data in the vehicle creates another major challenge regarding data privacy and security. With the power of connectivity also comes many possible vulnerabilities and threats, previously a cause of concern only in the information technology industry. Cyberattacks are becoming increasingly common on complex systems and have also reached the automotive industry. Attackers can use the vulnerabilities in the vehicle and steal the data or take control of the vehicle, which will affect the safety and security of users. Using your credit cards and bank details while making payments while sitting in the comfort of your vehicle gives you an added advantage. At the same time, the data you share in the vehicle can be accessed during any possible cyberattacks.

With all the advantages and benefits of SDCAVs, some challenges and concerns must be addressed. Adopting the famous Uncle Ben's dialog from the *Spider-Man* movie, "With great power comes great responsibility," we must remember this. With all the power of CPSs with connectivity and AI components integrated into the vehicle, vehicle manufacturers must ensure that these advanced technologies are deployed to provide a safe and secure user ecosystem.

3.8.
Summary

In the new technology world of the automotive industry, it is crucial to understand an SDV from a user's perspective. This chapter discussed what SDVs mean for the end user and how they can influence and impact their daily lives. Various ways in which vehicle functions can influence the user during the operation phase of the vehicle's lifecycle have been discussed. The story around software updates and the need to consider software updates in new vehicles have also been discussed in detail. The importance of OTA and how this is beneficial in bringing updates and upgrades in the vehicle was discussed in detail, as this will give a novice with better understanding of what OTA means and how it is beneficial for SDVs and any CPSs across different industries.

As providing the "WOW" factor was one of the key needs for software-defined approaches, vehicle manufacturers have adopted different ways to deliver that experience. The advantages and benefits of having the possibility to update the software and perform remote diagnostics have been discussed in this chapter. Predictive maintenance, an integral part of CPSs these days, helps create a better and safer driving environment and reduces repair costs. Knowing that a failure will occur will give the user a lot of advantage, rather than having a bad experience with the vehicle during the journey.

With all the advantages of SDCAVs, many challenges and concerns around them need to be addressed in the new era. As complexity and technology grow, negative aspects, such as challenges and concerns from an information technology perspective, are on the rise in the automotive industry. Addressing these concerns and challenges will create a safe and secure environment for users so that they can have an exciting experience during their journey.

References

3.1. Pathrose, P., *ADAS and Automated Driving: Systems Engineering* (Warrendale, PA: SAE International, 2024).

3.2. QNX, "Software-Defined Vehicles," The Ultimate Guide, accessed March 25, 2025, https://blackberry.qnx.com/en/ultimate-guides/software-defined-vehicle.

3.3. UNECE, "UN Regulation No. 155 - Cyber Security and Cyber Security Management System," accessed March 30, 2025, https://unece.org/transport/documents/2021/03/standards/un-regulation-no-155-cyber-security-and-cyber-security.

3.4. UNECE, "UN Regulation No. 156 - Software Update and Software Update Management System," accessed March 31, 2025, https://unece.org/transport/documents/2021/03/standards/un-regulation-no-156-software-update-and-software-update.

Chapter 04

Software Defined Vehicles: A Manufacturer's Viewpoint

The story of SDVs does not just end with how they influence and impact their end users daily. There is much more to the story. As we have briefly mentioned, SDV concepts are developed not just focusing on the benefits and advantages they could provide to the end user, but also including specific use cases and benefits for vehicle manufacturers. When looking at the complete lifecycle of the vehicle, many lifecycle phases have a direct relationship with the vehicle manufacturer.

This chapter will discuss how SDVs are seen from a vehicle manufacturer's perspective and how they can bring benefits and operational efficiencies to various processes in different vehicle lifecycle phases. When considering various lifecycle phases, ESs or other interacting systems that come into play should also be considered. Vehicle manufacturers utilize SDVs in different ways than what a user would assume.

For an automotive enthusiast, knowing about SDVs is very important in today's technology-driven market, where the automotive ecosystem is drastically changing. Rather than focusing on experiences from the end user's perspective, this chapter helps you gain insights into SDVs from the vehicle manufacturer's perspective, which is not typically visible to the external world. An individual who is just a vehicle buyer is usually unaware of the complex processes involved in the development and deployment of a vehicle in the market. It does not end there; the vehicle must be operational for many years, for which service and maintenance is required throughout its life, and it must comply with all the applicable regulatory frameworks during its lifetime.

A holistic view of how a vehicle is deployed in the market will help you understand how SDVs are conceptualized, built, and deployed, providing exciting experiences while using them.

4.1.
Manufacturer's World and Vehicle Lifecycle

When we buy a car with fantastic style and features, we are less concerned about how it is manufactured or the processes involved in getting such a vehicle to us. From a systems engineering perspective, a system or a vehicle (we consider it as the system here) must pass through various lifecycle phases until it reaches its end of life.

Over the various lifecycle phases of a vehicle, many stakeholders come into play and influence it. For an SDV, there are three crucial lifecycle phases where software can influence it and bring benefits and advantages to various stakeholders involved [4.1]:

1. Production phase
2. Operation phase
3. Service and maintenance phase

Two of these three lifecycle phases are dominated by the vehicle manufacturer: the production phase and the service and maintenance phase, as shown in **Figure 4.1**. In the operation phase, the vehicle has a higher influence and engagement from the user who owns it or the one who uses it. In addition, the operation phase of the vehicle is tightly bound to the service and maintenance phase to achieve extended life and to improve the quality and performance for its users.

SDV concepts and digitalization help vehicle manufacturers during the operation and service and maintenance phases of the vehicle lifecycle. Many other ESs or influencing systems are associated with these two phases, and the focus for a vehicle manufacturer is to improve the efficiency of the processes associated with these lifecycle phases, employing software and digitalization concepts using enablers such as AI, cloud data platforms, etc.

Figure 4.1 Manufacturer and vehicle lifecycle phases.

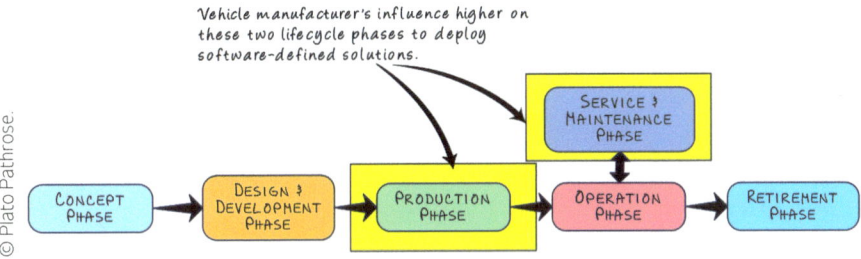

When we further investigate vehicle lifecycle phases, we can find out that multiple processes are involved in each phase. The production phase, one of the complex lifecycle phases, starts with mechanical and hardware processes and ends with the final integration and qualification, including the E/E components in the vehicle. Even though software can play a role in the production phase from software-defined concepts, these concepts are only applicable from the stage when various E/E modules are integrated into the subassembly or general assembly stages of the vehicle in its production process. One of the prominent areas where software-defined concepts, digitalization, and usage of enablers in the production phase of the vehicle are associated during the final assembly and quality control processes in production.

Figure 4.2 Will software fix my car?

Similarly, in the service and maintenance phase, it is understandable that software's influence on maintenance is limited to E/E components and their operations in the vehicle. This means we still need to take the vehicle to a service center when there is a critical issue, such as a mechanical breakdown or a hardware failure (**Figure 4.2**). However, in today's vehicles, in which complex electronic systems and software are integrated, a lot can be done for various vehicle functions by only concentrating on the electronic components and software, thanks to the technological advancements we have now. We will learn how it is done in the following sections, including case studies and applicational use cases.

4.2.
Why Should Manufacturers Utilize SDV Concepts?

When we discuss SDVs, we tend to approach them solely from the end-user perspective. However, many vehicle manufacturers have gone further and are utilizing software-defined concepts to improve their operational efficiencies in various processes associated with the production, service, and maintenance phases of the vehicle.

A vehicle manufacturer must depend on its suppliers for most of the E/E components in the vehicle. Even though the components are sourced from a Tier 1 supplier, many other suppliers at different levels, such as Tier 2, contract manufacturers, etc., are involved in developing and producing parts of these components. Software is also one of those components where different suppliers are involved.

When we look at the software ecosystem of the whole vehicle, it becomes transparent that it is a complex ecosystem. With complex electronic components and architecture, software dependencies also become complex. This reflects dependencies on various suppliers for the development, integration, and deployment of each component, making vehicle production complex. One can look at the BoM for hardware and software components in the vehicle to understand how many components and suppliers a vehicle manufacturer must deal with for the vehicle manufacturing process. There are many non-listed dependencies beyond the BoM list. As always, Tier 1 suppliers depend on Tier 2 suppliers, software providers, and so on to build each of these components.

With too many software components in the vehicle, it is challenging to coordinate them and make all of them available in the right quality with the correct set of functionalities on specific program gateways and targets. Those familiar with vehicle programs understand that running a vehicle program, considering the complex supply chain and deliverables from an enormous list of suppliers, is challenging while sticking to the plan.

In many vehicle programs, it becomes practically a challenge to deliver software to complex CPSs with all the functionalities in ideal quality and performance. These challenges are managed through different development models, such as agile, DevOps, etc. Nowadays, software updates have become part of these development models and are even extended to the production and operation phases, involving various electronic components in the vehicle.

Integration of various electronic components and their software elements is the most challenging part of the vehicle development process. Integration flaws are usually managed to a certain extent with software adaptation among the integrating components in the vehicle, which also drives software updates.

When specific flaws are identified later, i.e., once the vehicle is sold to a user, they can be managed with software updates if designed accordingly, unless these flaws are critical to fix with software adaptation and can affect the safety of the user. However, the critical use case of SDVs to engage and delight the user should not be forgotten. This can be achieved by providing features and services during the operation phase and employing software updates for new functions and improvements.

The primary benefit of applying software-defined concepts at the vehicle manufacturer level is it reduces operational costs by improving the efficiency of various processes associated with the vehicle's production and service and maintenance phases. At the same time, this also acts as a rescue mechanism for managing any future risks that may lead to vehicle recalls. If any flaws in the functionality or design are identified once the vehicle is sold and if the vehicle manufacturer can manage and fix them through software updates, it will save millions of dollars for the vehicle manufacturer by avoiding a possible recall of vehicles and fixing them through OTA software updates.

Software updates have become frequent and standard in product deployment with new development methodologies. Suppose that we have frequent software updates that extend to different vehicle lifecycle phases; then, it is critical to conceptualize and design the vehicle and its components to be compatible with these updates and strategies for fixes. We will discuss further in this chapter how different vehicle manufacturers plan to utilize software-defined concepts along with the digitalization and usage of technology enablers in their design to improve operational efficiencies and save costs in the overall vehicle development and deployment.

4.3.
What Can Be Done in Production?

The production phase of a vehicle can be viewed as a set of streamlined processes involving various components, the assembly process, and testing and qualification before moving to logistics. How can an SDS concept be used in a beneficial way in these processes?

The concept of SDVs can be applied in the production process where all electronic components are assembled and software is flashed. Many of the CPSs are shipped preflashed with software to the assembly line. These software versions may usually get upgraded or updated even when the vehicle is on the assembly line with an updated release from the supplier, either with an additional set of functions or to fix a bug and provide stable software to the system.

Having connectivity available in the facility or utilizing vehicle connectivity, any specific CPSs in the vehicle in an assembly line can be selected to manage these software updates. Using the latest digital infrastructures and assembly lines and deploying industry 4.0 concepts for smart manufacturing will facilitate rapid software updates and management even during specific processes associated with the production of vehicles in the assembly line (**Figure 4.3**).

Figure 4.3 Vehicle updates in an assembly line in a smart factory.

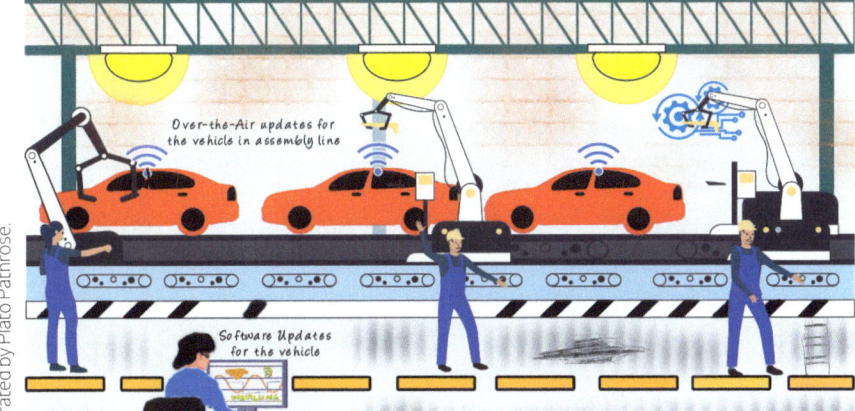

A few years ago, many vehicle manufacturers started considering autonomous driving functionalities for vehicles in their manufacturing facilities. One such case that hit the headlines was the BMW (Bayerische Motoren Werke) plant in Dingolfing [4.2]. Automated driving functionalities are integrated into one of the CPS suppliers' software and are activated to operate only in the factory environment using specific infrastructure-based sensors. These specific custom functions for production facilities, such as automated driving, might have been developed by either the vehicle manufacturer themselves or a supplier specifically for a certain ecosystem and are integrated into one of the CPSs of the vehicle to be active only during this phase of the vehicle lifecycle. These functions are usually deactivated once it moves out of the production facility.

How can these automated driving functions in the factory environment help the vehicle manufacturer? They help improve operational efficiency in factories with less manual involvement, extended operations, and optimized and continuous shifts from one process to another, thereby improving the takt time in production. Takt time is the rate at which the product must be completed in production to meet customer demand. For a vehicle manufacturing, it is calculated by dividing the available production time (net production time) by the customer demand (number of cars required).

Remote diagnostics and debugging are other tasks integrated into the vehicle in production facilities based on software-defined concepts. Diagnostics and debugging are standard procedures that cannot be skipped in any vehicle manufacturing process. They are part of the test and quality assurance process, where the finished vehicle is checked for issues or any warnings based on diagnostic flags from any electronic components in the vehicle.

Vehicles are also tested and evaluated for various functions to ensure functional correctness. Remote testing and diagnostics are commonly performed to facilitate quick transitions and checks from one vehicle to another in a production line, covering each and every vehicle without human interference or with minimal supervisory role by a human, thereby increasing operational efficiencies.

Calibrating external sensors such as cameras, radars, lidars, etc., has become a mandatory part of production for vehicles in which they are integrated. All these sensors act as essential elements for vehicles with ADAS and automated driving functionalities. Software-defined concepts and integrating software elements that can self-calibrate these sensors in the production and operation phases of the vehicle will help the manufacturer cut short the investments for expensive calibration stations in production facilities and service centers [4.1, 4.3].

All these processes discussed above are bits and pieces of the more extensive production process. Optimizing various processes will save money for the vehicle manufacturer and improve production efficiency. This can include savings in terms of human resources, tool licenses, and infrastructure at the factory. Introducing autonomous driving functionalities for moving the vehicle within the factory premises provides continuity in operations without depending on human resources to perform less productive activities over a longer period. This brings excellent advantages to the vehicle manufacturer in the vehicle production phase.

4.4.
What Can Be Done for the Service and Maintenance Phase of the Vehicle?

Once a vehicle is out of production and sold to a customer, the next most crucial phase where the manufacturer has a considerable influence is its service and maintenance phase. Usually, vehicle manufacturers partner with various dealers or establish direct sales and services to their customers. Service and maintenance is essential for keeping a vehicle safe, maintaining its performance, and extending its lifespan. Ultimately, this means saving money on costly repairs and keeping the vehicle in good condition for a longer period. How can SDVs help vehicle manufacturers in this lifecycle phase?

From the systems engineering viewpoint, the service and maintenance phase of the vehicle works along with its operation phase. These two phases are tightly coupled, and any failure in providing a good service and maintenance experience can affect the customer's interest in the brand.

Vehicle manufacturers have come up with different solutions using SDV concepts to improve the way service and maintenance is offered and utilized, which is an opportunity to improve customer engagement with their vehicle and the brand. One such solution is the remote service concept. With this concept, the user is not required to take the vehicle to the service station. Instead, the required services and software updates for vehicle components can be done remotely using high-speed connectivity provided in the vehicle [4.3, 4.4]. This could be possible using a Wi-Fi network or other network options such as 5G/4G/3G integrated as part of vehicle connectivity.

Traditional ICE vehicles have more moving parts, hence requiring refilling and replacing oils and lubricants. These days, with the transformation to EVs, it is said that there are approximately at least 2000 fewer moving parts than in traditional ICE vehicles. This also reduces the need for oil changes and lubricants, thereby changing these vehicles into mere electronic devices that operate with CPSs integrated with high processing power, storage, and connectivity.

Previously, software updates in the vehicle for bug fixes, enabling new functions, and identifying a warning displayed required the user to visit service centers. All these are now done in the comfort of the garage or even on the streets. This helps vehicle manufacturers to deploy software to the vehicle in a more strategic way. The time to take the vehicle to the service center for software updates to fix certain electronic malfunctions has been drastically reduced. This has dramatically changed the way software updates and bug fixes are deployed in the vehicle (**Figure 4.4**).

Figure 4.4 Remote diagnostics and repair.

Having frequent software updates that fix bugs and improve the performance of various systems in the vehicle and their functionalities delights the customer. Even remote connectivity to the vehicle manufacturer's data platform or server helps the vehicle manufacturer identify the issues in the vehicle and its current health status. The flow of information from various sensors integrated with the vehicle about the health status of various components can be utilized to analyze and identify what is going wrong and how it will impact the usage of the vehicle. Utilizing CPSs with AI algorithms operating on these data helps identify and warn the user as required. Once a probable issue is identified, a solution can be planned through software updates or informing the driver to act accordingly so that it does not affect their journey due to a broken vehicle.

This remote analysis and predictive maintenance is a blessing in today's world, as lifestyle has become busy; having a broken vehicle while in use will seriously affect one's life. So, vehicle manufacturers have started utilizing SDSs to help improve the quality of life of their customers.

With the transformation of vehicles and the modular manufacturing of various vehicle parts, one may not be required to visit a service or repair center to get certain parts repaired or replaced in the future [4.3, 4.5]. The defect can be remotely identified, and the corresponding replacement parts will be shipped to the user. Once the parts are shipped and delivered, the user can assemble them with remote support and configuration that has been made possible via the vehicle manufacturer's technical support team (**Figure 4.5**). This can even eliminate the role of dealers from the vehicle sales and service ecosystem.

Figure 4.5 Vehicle repair and remote support in the future.

The future of remotely performing repairs and replacements in vehicles is not so far, as we have already seen many vehicle manufacturers avoiding dealers and even selling vehicles directly to customers. This business model also reduces the need to establish dealer based service centers for their vehicles. Thus, this software-defined concept in vehicles provides many benefits for vehicle manufacturers.

4.5.
Changing Ecosystem of Regulatory Frameworks

In the above sections, we have discussed the benefits that a vehicle manufacturer will have with SDVs during their production, service, and maintenance phases. In this section, we will discuss another maintenance use case that requires a vehicle to comply with the regulations of various geographical regions.

Compliance with regulatory frameworks determines whether a vehicle can be sold in a particular geographical region. These regulations are regularly amended and updated to control various new technologies that are integrated into vehicles and to improve users' safety while using them.

It is extremely challenging for a vehicle manufacturer to remain compliant with the regulatory needs of various geographical regions while planning a vehicle program. A vehicle program is initially planned to target specific regions and regulations applicable to these regions. Changes in regulation can be foreseen via prior announcements. But things can go wrong, and vehicle programs can be delayed.

Delays in the deployment of the vehicle in the target market can be attributable to any of its critical components or a late identification of a problem in its core component. There can be many organizational reasons for such delays. These delays will impact the investment and the planned vehicle program if they result in adopting new regulatory needs or requirements when the vehicle is ready to be deployed in the market.

During these challenging times, manufacturers can integrate new software packages and address new regulatory requirements through software updates and upgrades to SDVs. This also gives vehicle manufacturers flexibility in targeting newer markets that were not considered when the vehicle program was initially planned. A different set of requirements beyond the existing ones can be well managed with software and minimal changes to the vehicle, as required for newer markets. This benefits vehicle manufacturers by allowing them to scale their market and expand into new regions with fewer hurdles for their products. SDVs have a risk mitigation mechanism for being prepared for the change and for addressing the ambiguities that may occur in the future during their deployment in various markets.

One typical example is the announcement of the UNECE on the adoption of GSR 2 requirements [4.6]. These announcements made many ADAS functions mandatory in vehicles, including the intelligent speed assist (ISA) functionality. The ISA system informs, warns, and discourages the driver from exceeding the statutory local speed limit. The vehicle sets its speed limit automatically based on the speed limits indicated on the road. The front camera of the vehicle and GPS, coupled with digital maps with speed limit information, allow the ISA technology to continuously update the vehicle speed limit based on the defined road speed limit [4.7].

Previously, complying with this new regulation was difficult for many vehicle manufacturers who were unprepared and those who manufactured vehicles with platforms that were not ready to accommodate additional electronic components. Some of the vehicle manufacturers who had an advanced platform and those who started utilizing CPSs as part of infotainment or ADAS in the vehicle were able to manage the introduction of ISA functionality more easily with fewer difficulties and engineering expenses by working on solutions together with different map suppliers. However, those vehicle manufacturers who were not prepared and did not have any CPSs were forced to address this issue by initiating this process on new vehicles and electrical platforms for their vehicles, which added tremendous engineering expenses due to these regulatory changes.

This gives an important lesson about preparing for the future by adopting new technologies, enriching ecosystems, and anticipating regulatory changes for specific markets. SDVs are the future, which will act as a risk mitigation method for vehicle manufacturers to address the ambiguities in the market and reduce costs on processes such as functional issues, where the vehicle manufacturer becomes responsible for vehicle recalls, if any.

4.6.
How Do SDVs Improve Efficiency?

SDVs integrate CPSs, which bring many advantages to the vehicle manufacturer in various lifecycle phases of the vehicle. We have discussed a few advantages in the above sections. A novice can think of SDVs similar to the smartphone that they hold in their hands. The automotive industry is looking forward to making users think of vehicles as smartphones on wheels.

The primary advantage of SDVs, beyond their huge dependence on software, is their ability to connect to the external network of infrastructures and devices. Since an SDV and its various CPSs depend on software for their complex functions, this connectivity empowers it with the iterative power of software updates for functional improvements and quality, even upgrading it by adding new features.

Let us focus on software updates in SDVs in their different lifecycle phases. Recently, the speed at which software updates are done has increased tremendously. Earlier, software updates to electronic components in the vehicle were made manually by technicians by physically plugging in cables to the vehicle network and flashing software. This has changed now with capabilities where a fleet of vehicles can undergo software updates by remotely managing them (**Figure 4.6**). This gives the power to do more in less time and using fewer resources.

Figure 4.6 Software updates for a fleet of vehicles.

Different possibilities for software updates are available to the vehicle manufacturer during the vehicle's lifecycle phases. The flexibility for the vehicle manufacturer and suppliers to manage frequent software updates can be managed even before the vehicle is rolled out of the production facility. Those who are familiar with vehicle manufacturing can understand the complexity of managing and coordinating software updates from various suppliers during the production phase of a vehicle program.

In earlier days, missing the timeline for the delivery of software could cause the vehicle program to skip those functionalities in the vehicle. It could even lead to a situation where the scope of the supplier is reduced since the timeline was not met. It could even affect the payment for the suppliers associated with the project. SDVs help manage the challenges of delayed software deliveries due to quality issues or technological constraints and make the vehicle an open platform to accept and manage the software at any time for integrated components.

Imagine a fleet of vehicles in the factory getting software updates all at once after production before they are shipped to various dealers. If the vehicles are transported via ships, they can receive software updates while they are in the harbor or even while they are on ships during their transit. These are not just imaginations anymore; they are real and have been practiced nowadays by various vehicle manufacturers. This is the advantage and flexibility SDVs bring due to their high software dependency and power to connect to the outside world.

This provides the vehicle manufacturer with flexibility regarding when and where software updates can be rolled out. This flexibility brings with it the power of change and effort to build better products with iterative approaches in no time.

4.7. SDVs to Save Vehicle Costs?

With all vehicle manufacturers concentrating their efforts to create SDVs, the benefits of these efforts are also extended to different areas beyond the product, such as savings in overall engineering and product costs. How are they benefiting the vehicle manufacturer and their customers in terms of costs and expenses?

Introducing complex CPSs with high-end processors that hold complex software, including AI elements, adds to the overall cost of the vehicle. The engineering expense for a vehicle manufacturer to integrate all such elements into the vehicle is also an additional expense that was not there earlier. With software-defined concepts utilizing connectivity to act as the backbone of modern vehicles, many vehicle manufacturers have thought about reducing the number of hardware components in the vehicle and taking all these high-end processing elements outside the vehicle, thereby reducing the cost of the vehicle.

The connectivity of vehicles plays a significant role here. Information from sensors is transferred to the outside world, where it is processed and analyzed and an appropriate action is defined. The information on how to execute the action is then transferred back to the vehicle, which then performs the action using its controls and actuators. Various European vehicle manufacturers have successfully implemented prototypes of certain low-speed ADAS and automated driving functions, such as automated parking and valet parking, in this method.

This reduces the need for high-end processors integrated into the vehicle system; instead, all processing tasks are transferred to the vehicle manufacturer's backend server in a cloud data platform. High-speed connectivity plays a significant role here. Functional software updates that analyze the data are only required at the data platform and are not directly deployed on each vehicle, which adds further flexibility and reduces engineering expenses for the vehicle manufacturer.

These approaches offer solutions to keep the existing or outdated E/E architecture or vehicle platforms operational for an extended period for many vehicle manufacturers. Especially for large-scale manufacturers, changing the vehicle platform and their E/E architecture is tiresome, expensive, and complex when they must continue producing different platforms in different vehicle models.

Another approach is replacing certain physical sensors with virtual sensors in the vehicle. This approach utilizes the capability of neural networks to provide accurate measurements, according to which various vehicle components can act. This has been used recently by a few vehicle manufacturers to replace their hardware sensors, such as vehicle mass sensors to measure the weight of the vehicle, brake fluid pressure sensors, etc. Using virtual sensors results in an advantage of reducing the hardware costs. The sensors and related components can be completely removed, which also provides increased supply chain flexibility. Virtual sensors, being a software component, can be integrated into any electronic component of the vehicle, such as a domain controller or zonal controller.

With this approach, there are significant advantages such as reductions in engineering costs, infrastructure costs, and the overall cost of the vehicle. However, this comes with a few dependencies and challenges. While deploying functional software remotely using cloud platforms, the connectivity of the vehicle acts as the central pillar. Having a low-speed connection or no connection will seriously affect the availability of these functions or have an impact on their performance in the vehicle. This will also require a subscription to high-speed internet connectivity in the vehicle, which the customer usually has to pay for out of pocket, like a smartphone internet contract.

4.8. Use Cases of SDCAVs

Knowing more about SDVs brings new ideas and possibilities regarding how software influences a vehicle's various lifecycle phases. For someone starting their journey on the engineering side of SDVs, it feels pretty challenging to identify a list of specific use cases based on the limited knowledge and information about the vehicle and its internal architecture.

Without looking at the vehicle internals, let us take an artist's mindset, knowing various available functions in the vehicle that can be improved for its efficiency with the help of software and, of course, the power of connectivity in the vehicle. As we all know, many vehicle manufacturers are still utilizing a hybrid architecture with a mix of traditional electronic components and new CPSs in their vehicles. Entirely replacing the E/E components in the vehicle to establish a new platform will be expensive. This will add further challenges around the supply chain, contracts, and guarantee topics with many new and existing suppliers. Hence, it is a reality that most traditional and emerging vehicle manufacturers have two parts in their architecture: a traditional set of components and the latest CPSs. We will discuss this later in this book when analyzing various E/E architecture aspects in SDVs.

Considering all the existing functions of the vehicle, one can identify the use cases by decomposing the vehicle's lifecycle into various phases. A brief overview of various use cases for the three lifecycle phases of a vehicle is presented in **Table 4.1**. Whether it is the production and service and maintenance phases, where the vehicle manufacturer has higher influence, or the operation phase, where the vehicle is in use, these use cases act as fundamentals for any solutions targeting SDVs.

Table 4.1 Use cases for SDCAVs.

SDCAV lifecycle phases and various use cases	
Production phase	
1	Vehicle configuration and variant coding of various modules in the vehicle.
2	Feature calibration of various vehicle features and associated functions.
3	Automatic sensor calibration for cameras, radars, lidar, etc., that has to be done at the end-of-line (EoL) in production.
4	Self-tests and diagnostics of the vehicle and vehicle components.
5	Automated driving within the production facility (autonomous or supervised driving).
6	Software flashing and updates of various components in the vehicle (flashing in the production line, at the storage area, or during logistics).
7	Using virtual sensors in place of physical sensors and harnesses (e.g., mass sensors, brake pressure sensor, etc.).
Operation phase	
1	Software updates for various vehicle components. Wired and OTA updates for bug fixing, performance improvement, optimization, etc.
2	New feature addition (wired and wireless deployment). It could also be feature-on-demand (FoD) or mandatory deployment for regulatory compliance.
3	Subscription-based services to users (payment services, music, streaming services, games, etc.).
4	Automatic vehicle health and status checks and reporting problems to the service center.
5	Proactive issue identification and information services to the user.
6	Establishing connectivity to various infrastructures and platforms for specific use cases.
7	Personalization of various vehicle functions and third-party services.
8	Remote execution of certain vehicle functions (low-speed parking, pet mode, summon mode, etc.).
9	Providing shared access to the user based on their profile with different platforms (smart home services inside the vehicle and home devices such as television, smartphones, etc.).

(*Continued*)

Table 4.1 (Continued) Use cases for SDCAVs.

SDCAV lifecycle phases and various use cases	
10	Executing various safety and comfort functions in the vehicle, with communication between other vehicles and infrastructures in various driving conditions.
11	Value-added services from the vehicle manufacturer based on data analytics of driving style and travel patterns.
12	Specific connected services for various vehicle-level use cases such as e-call, s-call, etc.
13	Fleet management, supervised guidance, and route allocation.
14	Remote driving and support services.
15	Location-based services and third-party services.
16	Function optimization with live status and vehicle performance tracking for better mileage and eco-driving.
Service and maintenance phase	
1	Remote diagnostics and issue analysis for the entire vehicle.
2	Predictive maintenance and service of the vehicle.
3	Remote software update and bug fixes on demand or specific to a particular release.
4	Software updates, subscription renewal for services, service-level agreements, and try-outs of new features.
5	Vehicle road worthiness status, service history, and tracking.
6	Vehicle health monitoring and service scheduling.
7	On-demand support from the customer support team.
8	Deployment of tailored solutions.

Table 4.1 presents an overview of various use cases and concepts to build software-driven solutions around different functions available at the start of the journey of an SDV. The use cases on how SDVs influence the user and how various functions can be adapted during the vehicle's operation phase were discussed in Chapter 3, where the customer's viewpoint on SDVs were presented.

In SDVs, use cases are fundamental for identifying and investing in the technology. Thus, knowing which problems can be solved, which functions can be improved, and how they will serve the user and the vehicle manufacturer is critical. Investing in technologies for SDVs without identifying the purpose or use case will not be beneficial. Today, we can see that many technologies are integrated into vehicles that the user may not need or might not even use over their lifetime.

4.9. Challenges and Restrictions

In this chapter, so far, we have discussed the advantages and benefits of SDVs. However, there is always another side to the story, and we must accept and be aware of the challenges and disadvantages, apart from numerous benefits.

We have discussed three technology areas for SDVs: software, which allows for changing anything on the CPSs integrated in the vehicle, making it dynamic over its lifetime; connectivity, which allows the vehicle to be connected to the external world; and autonomous functionalities, which can improve the operational efficiency of the vehicle or can be provided as an additional feature to the end user. The major challenges for SDCAVs can also be described around these three areas [4.3, 4.4, 4.5], as shown in **Figure 4.7**.

Figure 4.7 Challenges and restrictions in deploying SDCAVs.

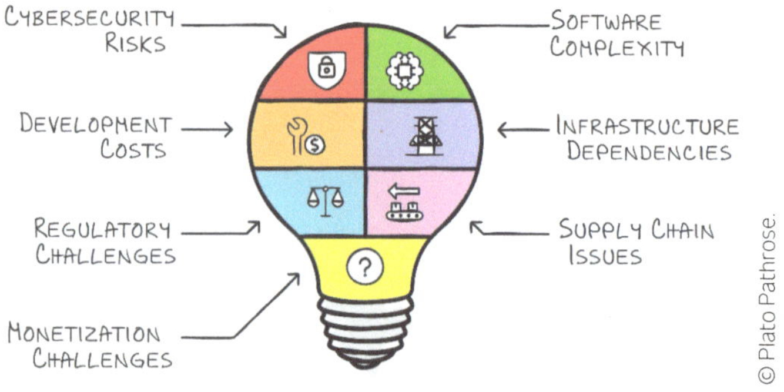

1. Cybersecurity risks:

    The increase in connectivity and dependencies on software for various components in the vehicle increases the risk of cyberthreats. One can tap a lot of personal and financial information of a user from the vehicle if the user uses the vehicle for various applications such as shopping, reading messages, pairing phones to the vehicle, loading personal details, etc. These random exploits and attacks can even affect the safety of the vehicle and can also be a threat to the privacy of the user.

2. Increase in software complexity and reliability:

   With complex CPSs being part of SDVs, dependencies on software and its quality are a major challenge. Having safety-critical functions operating from CPSs, such as addressing software bugs that influence their functions, and managing the complex software in CPSs must be done cautiously. The flexibility of software updates without any dependency on time and place has also made many vehicle manufacturers focus on getting their product to the market as quickly as possible, rather than providing a quality and safe product. This mindset is dangerous, especially with SDCAVs.

3. Higher cost of development and maintenance:

   Due to the complex software that comes along with these complex CPSs and various functions in the vehicle, investments in the development of both software and hardware have become enormous. Especially when integrating AI-based functions, a lot of efforts and investments are required during their development and deployment cycle. Continuous monitoring and maintenance of software over the vehicle's lifecycle adds further expense to the manufacturer.

4. Infrastructure and connectivity dependencies:

   With an SDV, the major technology dependency is software and connectivity, which requires specific contracts and agreements. Sometimes, this adds to the customer's cost when they have different subscription models, payments for cloud platforms, and internet charges for the vehicle.

5. Regulatory and compliance challenges:

   Regulatory frameworks are evolving around new technologies, such as the usage of AI-based software, data privacy, and autonomous functionality. The vehicle manufacturer needs to seriously follow up on the changing regulations and be prepared with appropriate solutions to comply with them, which can be challenging. However, this becomes an advantage for the user as it will result in having a longer lifetime for the vehicle than it would have been without connectivity and software-driven systems.

6. Supply chain dependencies:

   One of the significant challenges for vehicle manufacturers in building SDVs is their supply chains. Dependency on certain suppliers for high-end processors will increase the pricing of components and result in bottlenecks on deliveries, as well as delaying the delivery of these components. This is a significant challenge when there are fewer suppliers for a specific component of a product with high demand. Such components include high-end microprocessors for ADAS and autonomous driving, infotainment systems, connectivity modules, etc.

7. Monetization and return on investments (RoI):

   Even though SDVs have numerous benefits and advantages, it is a very challenging ecosystem for vehicle manufacturers to harness and build a business around them. Customers are always cost-sensitive. They hate repetitive payments for the vehicle and investing further in it after the purchase. It is challenging for vehicle manufacturers to convince customers and make them subscribe to the features that they offer in the vehicle. For instance, across different vehicle manufacturers, many customers do not renew their subscriptions to the features and services provided in the vehicle after they expire. It is always a question of whether the returns are enough and whether a successful business model can be established with the subscription model alone, considering the investments in these technologies by the vehicle manufacturer.

4.10. Summary

The vehicle manufacturer is a major player in the ecosystem of SDVs. This chapter discussed the vehicle manufacturer's viewpoint on how SDVs can be beneficial to them and help them build their business. The production and service and maintenance phases are the two lifecycle phases of a vehicle where the manufacturer has a strong influence. We have also discussed what SDVs can do in these two lifecycle phases and how they benefit the manufacturer. In addition, we reviewed the significant impact on the user during the vehicle's operation phase.

The flexibility SDVs give to vehicle manufacturers and end users is excellent. SDVs are a good example of how technology is driving toward making our vehicles look like smartphones on wheels. Apart from these benefits, we have also discussed the challenges and constraints vehicle manufacturers face while building SDVs or integrating CPSs into them.

Even though a lot of features are available in SDVs, many vehicle manufacturers face the question of how many times a user will use these features while in the vehicle. This has led many vehicle manufacturers to follow subscription-based business models. With customers being cost-sensitive, it is demanding and challenging for vehicle manufacturers to find the sweet spot in establishing a better business model to build a business around it. This is one of the primary reasons why customers do not renew their subscriptions for many features that are available during the purchase. Let us watch out for the market, evolving technologies, and how different business models emerge from various players in the journey of SDCAVs in the upcoming years.

References

4.1. IEEE SA, "ISO/IEC/IEEE 15288:2015 Systems and Software Engineering—System Life Cycle Processes," 2015.

4.2. BMW Group PressClub, "High-Tech in Production: BMW Group Enables Automated Driving for New Vehicles," accessed March 27, 2025, https://www.press.bmwgroup.com/global/article/detail/T0446493EN/high-tech-in-production:-bmw-group-enables-automated-driving-for-new-vehicles?language=en.

4.3. Pathrose, P., *ADAS and Automated Driving: Systems Engineering* (Warrendale, PA: SAE International, 2024).

4.4. Lopez-Herrejon, R.E., Martinez, J., Assunção, W.K.G., Ziadi, T. et al., *Handbook of Re-Engineering Software Intensive Systems into Software Product Lines* (Cham: Springer, 2023).

4.5. QNX, "Software-Defined Vehicles," The Ultimate Guide, accessed March 25, 2025, https://blackberry.qnx.com/en/ultimate-guides/software-defined-vehicle.

4.6. European Commission, "New Rules to Improve Road Safety and Enable Fully Driverless Vehicles in the EU," accessed March 28, 2025, https://ec.europa.eu/commission/presscorner/detail/en/ip_22_4312.

4.7. European Commission, "Intelligent Speed Adaptation (ISA)," Mobility & Transport - Road Safety, accessed March 29, 2025, https://road-safety.transport.ec.europa.eu/european-road-safety-observatory/statistics-and-analysis-archive/esafety/intelligent-speed-adaptation-isa_en.

Chapter | 05

Software Defined Connected and Autonomous Vehicles: An Architecture Overview

Individuals with an engineering background find the architecture of any cutting-edge product as one of the most important and interesting areas. The concept of SDVs is one of the most discussed things in the mobility industry nowadays, so it is interesting to have a deeper look at SDVs and understand how they are conceptualized, designed, and built. This chapter is intended to present an overview of the evolution of automotive technology and the architecture of SDVs from traditional approaches.

Architecture does not just cover CPSs or SDSs in a vehicle. Instead, it encompasses the entire vehicle to understand how software will add value to the user and the vehicle manufacturer. A systems engineering approach is followed in this chapter to describe and clarify the architectural evolution and approaches in establishing SDCAVs. This will help readers from a non-engineering background understand the subject and visualize SDVs and how they will impact the lives of millions of people using them.

5.1. Vehicle-Level Architecture of an SDCAV

One of the most critical areas in the development of a product is its architecture. Since SDCAVs have become the backbone of the mobility industry, it is interesting to know how they are designed and what they consist of. As we know, many legacy vehicle manufacturers struggle to compete with the new and emerging players in the automotive industry due to technological transformation and advancements in production timelines. Thus, understanding how these legacy manufacturers are trying to adapt the technology along with their existing infrastructure and ecosystem is essential.

An entire vehicle can be decomposed into various subdomains and subsystems, as discussed in Section 2.2 of Chapter 2. This is a function-based decomposition covering various vehicle-level functions and modules that facilitate these functions, and it will give you a high-level overview of decomposition from an SoS perspective. We have also learned that any of these systems can be built as a software-defined CPS that can be further scaled into SDVs [5.1].

Here, to provide a better understanding to non-engineering readers, we will present a logical decomposition of the vehicle into different layers. **Figure 5.1** shows a five-layer architecture based on the logical decomposition of an SDCAV.

Figure 5.1 Architecture of SDCAVs.

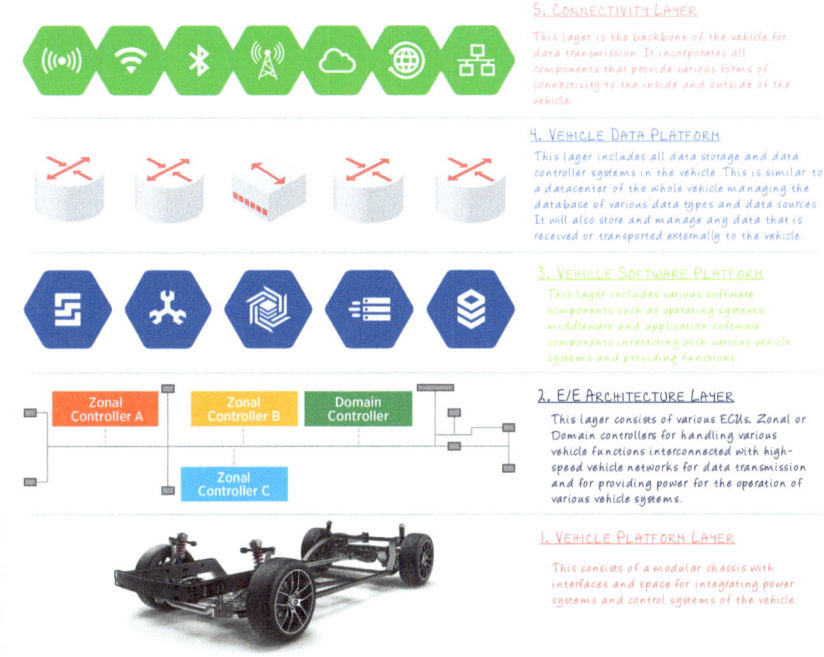

This five-layer logical architecture considers the concept of hardware and software independence. This is one way in which SDVs can be differentiated from traditional vehicles. It provides greater flexibility in adapting the functions of the vehicle through software without depending on hardware components integrated into it.

In this bottom-up decomposition, the first layer is the vehicle platform layer, which is the foundation for any vehicle. Vehicle manufacturers use various modular vehicle platforms these days. With the adoption of electrical technology as the main power source for vehicles, a modular chassis with interfaces to integrate battery packs and control devices, such as actuators, serves as an optimal platform for any vehicle.

The second layer is the E/E architecture layer, the heart of an SDCAV. This layer incorporates various CPSs and all the associated E/E modules to help them operate efficiently. It can range from a mix of electronic devices such as ECUs, domain controllers, zonal controllers, switches, etc., to various harnesses connecting these systems. The key element in the E/E architecture of the vehicle is the harness that connects various electronic devices. This includes cables to power these devices and the network for data exchange among these devices and various sensors integrated into the vehicle.

Usually, ethernet, controller area network (CAN), and local interconnect network (LIN) are parts of the vehicle network that allow data and information exchange among various devices. They are well-known protocols that allow information exchange through cables connected across different E/E architecture components. Wireless connections are also established for data exchange between various components in the E/E architecture. This is the layer where any CPS or SDS will have its entry into the vehicle. All the layers above the E/E architecture layer are closely bound to and strongly depend on this layer, as their architecture and deployment depend on CPSs and components integrated into the E/E architecture. However, as discussed in previous chapters, there is no need for an SDCAV to be an EV. The classification of whether an SDV is an ICE vehicle or a battery electric vehicle (BEV) is possible only when we look at the driving power source behind these vehicles.

The third layer is the vehicle software platform layer. This layer strongly depends on CPSs deployed in the E/E architecture layer. Many vehicle manufacturers still operate with hybrid E/E architectures, which integrate multiple ECUs, domain controllers, and zonal controllers along with other components. Software layering and deployment of various CPSs and their information exchange depend on the layers below. The software platform layer usually incorporates multiple OSs, middleware for high-speed data exchanges, and application software for functions and interfaces. It acts as a bridge for data exchange among various E/E components via the vehicle network.

Above the software platform layer lies the vehicle data platform layer. This layer integrates electronic components intended for data storage and management. Depending on the complexity and functions of the vehicle, this can range from a simple storage disk in the CPS to an event data recorder (EDR) and data storage system for autonomous driving (DSSAD). With many sensors and the complexity of the software and functions together with the connectivity that SDCAVs bring to the vehicles these days, the vehicle data platform layer has an important role to play.

The vehicle data platform layer plays a crucial role in the generation of data from various sensors and CPSs in the vehicle, the exchange of data between external infrastructure and platform, the provision of data for remote diagnostics and predictive maintenance, the collection and management of personal data following regulatory frameworks, and security, among other things.

With the new generation of vehicles becoming highly software-centric and acting as data centers, collecting and processing vast amounts of data, the importance of the vehicle data platform layer has significantly increased. AI usually manages and processes a vast amount of data to avoid further software and hardware complexity. Multiple storage and memory modules are present in new generation vehicles, which shows the increasing importance of the vehicle data platform layer.

The topmost layer in vehicle architecture is the vehicle connectivity layer. This layer is the window of SDCAVs to the external world. Even though it manages specific wireless connectivity within the vehicle, it is mostly the communication interface of the vehicle to all the stakeholders outside the vehicle. The vehicle connectivity layer manages data flow and information exchange from the vehicle to external stakeholders, such as external devices, infrastructure, vehicles, data platforms, etc. All these interfaces provide the possibility to exchange data and software, thereby managing software updates, information exchanges, and entertainment, and providing a better experience to the user during the journey.

The vehicle connectivity layer provides protocol-based wireless connections such as Wi-Fi, 5G/4G/3G, Bluetooth, Whitefire, wired connections, ethernet, USB, and CAN to external devices and platforms. It also manages the information flow for certain functions available in the vehicle, such as live streaming of internet data, like movies, online gaming, etc. Many of these remote features available in the data platform or the data server of the vehicle manufacturer are based on the vehicle connectivity layer in SDCAVs. Without this layer, having fancy features and communicating with the external world becomes challenging or almost impossible.

5.2.
Evolution and Applications of E/E Architecture

We have learned that in the layered logical architecture of SDCAVs, the E/E architecture layer plays a critical role in defining various layers and their capabilities and making the vehicle software-defined. The E/E architecture in an SDV can be compared to the skeleton in a human body, and the software acts as the flesh around the skeleton, providing features and different functions [5.1, 5.2].

Learning about the evolution of vehicle architecture is significant to understand the current technologies. **Figure 5.2** presents an overview of various vehicle architectures used by vehicle manufacturers in the market, from distributed to zonal architectures.

Figure 5.2 Distributed, domain-centralized, and zonal architectures.

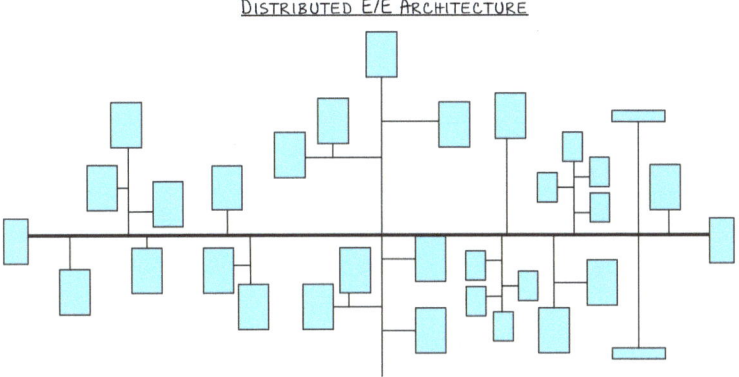

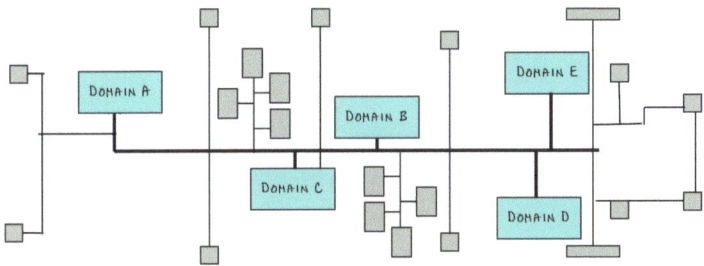

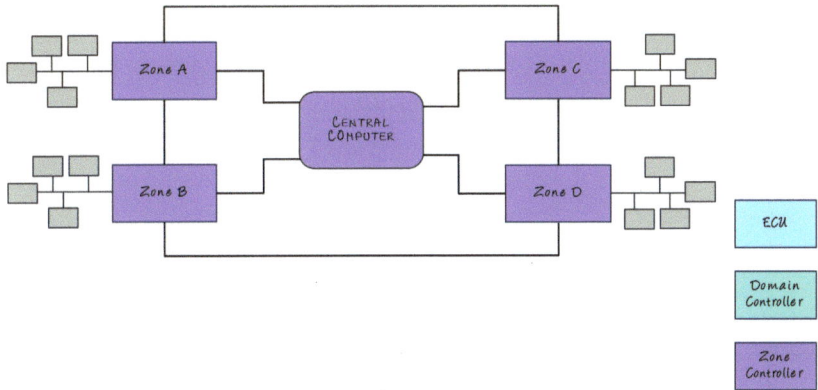

5.2.1.
Distributed Architecture

Earlier, the E/E architecture of a vehicle was distributed in structure with multiple ECUs, and each ECU was targeted to address specific vehicle functions. This structure was too modular, and each function within the vehicle was associated with one or more ECUs operating together. This increased the complexity of managing more components, and the harness associated with the communication between these ECUs became a challenge. Moreover, having a lot of nodes in the E/E architecture increased power consumption and power wastage through heat generation. Another significant challenge with this type of architecture was that it required more electronic components and connectors, thereby increasing the number of suppliers and making the integration complex. It also increased the harness length, thereby increasing the overall weight of the vehicle.

5.2.2.
Domain-Centralized Architecture

The domain-centralized architecture was an improvement from the distributed architecture, which addressed the consolidation of various ECUs by introducing domain controllers for specific functions associated with each domain. It is a concept of decomposing and classifying overall vehicle functions into various domain-specific functions.

Domain controllers consist of one or more microprocessors operating on the same or different OSs with a collection of software components that address the functions associated with a specific domain area in the vehicle. Infotainment domain controllers, ADAS domain controllers, and gateway domain controllers are examples of such domain controllers.

5.2.3.
Zonal Architecture

Zonal architecture, or integrating zonal controllers, results from the decomposition and classification of vehicle functions into various zones in the vehicle. Zonal architecture uses zonal controllers with almost similar classification to domain-centralized architecture. The main difference between domain controllers and zonal controllers is their processing power and the functions they can integrate. Zonal controllers, with additional processing power, can integrate heavy AI algorithms to provide new functions and improve the performance of various vehicle functions. These high-performance computers are also centers for managing massive amounts of data from sensors and other components within and outside the vehicle. Based on the configuration of zonal controllers, they can act as a master that controls multiple ECUs or domain controllers, which address various functions for specific zones in the vehicle.

Unlike ECUs from distributed architecture systems, domain and zonal architectures use fewer and lighter harnesses. Because of their higher bandwidth, these harnesses can carry a huge amount of data in no time. Gigabit Ethernet (GigE) interfaces are the most commonly used interfaces for connecting these CPSs to meet the data bandwidth requirements in zonal architectures.

5.2.4.
Hybrid Architecture

There are numerous advantages in utilizing domain-centralized and zonal architectures in SDVs. However, when we dive deep into the current automotive industry, we find that many emerging vehicle manufacturers use domain-centralized and zonal architectures for their E/E architecture to a higher extent than legacy automotive manufacturers, which sell millions of vehicles yearly. These legacy vehicle manufacturers are moving ahead with hybrid vehicle architecture, which is a mix of distributed, domain-centralized, and zonal architectures.

Legacy manufacturers use these hybrid E/E architectures because of the dependencies with many legacy elements in these vehicles and the volume of vehicles the manufacturer should consider changing if they need to switch to a complete zonal or domain-centralized architecture. Many organizational reasons also contribute to this delay of legacy vehicle manufacturers in adopting a complete domain or zonal architecture. **Figure 5.3** shows a hybrid architecture that is still popular among legacy vehicle manufacturers.

Figure 5.3 A hybrid E/E architecture.

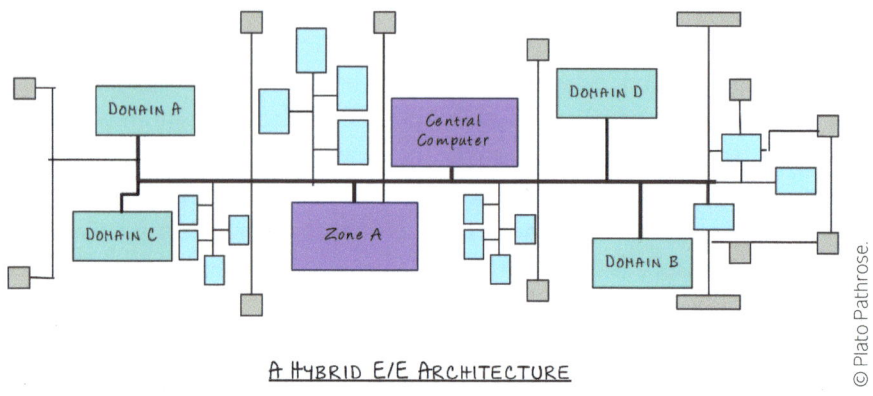

When we look at many of today's SDCAVs, we find that the majority of them use a hybrid architecture. This is mainly because of the following reasons:

1. A domain controller or zonal controller may not be available for certain functions from the qualified supplier within the supply chain.
2. There could be a legacy dependency on the vehicle platform or E/E architecture that prevents the adoption of zonal controllers in the E/E architecture.
3. Building a new E/E architecture and platform adds enormous cost to the vehicle's engineering development.
4. Adding domain and zonal controllers will automatically increase the vehicle's BoM cost, which customers see as overpriced.
5. The additional value that the introduction of zonal or domain-centralized architectures would bring is limited by the noncompliant platform used in the vehicle.
6. There might be delays in production-ready components from the supplier after qualification and integration challenges with other components in the vehicle.

All these factors influence legacy vehicle manufacturers to use the hybrid E/E architecture in their vehicles. Usually, a hybrid architecture integrates certain domain controllers or zonal controllers, dedicated to a specific domain or zone, and the rest of the E/E architecture still utilizes legacy components, such as ECUs, which were part of their earlier distributed architecture. This is usually partitioned based on different network channels in the vehicle's E/E architecture. One or two network channels hold advanced CPSs with high-speed networks such as GigE; on the other hand, other network channels might still use CAN interfaces for communication and data exchange. From a user perspective, an SDV can be used with any E/E architecture; the only thing that matters is how the software brings value to the user while these systems operate together inside the vehicle.

5.2.5.
The Need for a Platform Approach in E/E Architecture

Large-scale vehicle manufacturers that produce millions of vehicles every year face huge challenges in transforming to SDCAVs. Most of those are driven by their dependency on legacy platforms and infrastructures. This is where emerging vehicle manufacturers have an advantage: they do not have many dependencies and are flexible with their development processes and how they build vehicles. Moreover, they are more agile for organizational reasons, enabling them to introduce advanced technologies and take paths that are uncommon among legacy manufacturers.

With the scaling up of vehicle models and numbers to address deployments in various geographical regions, it is always better to move ahead with a platform approach for the E/E architecture. One of the most common mistakes made by vehicle manufacturers when they introduce domain and zonal controllers in their vehicles is these controllers are integrated with harness and networks that do not have enough data bandwidth. The networks that connect them are not high speed, which limits the ability to use these domain or zone controllers to their full capacity. This limits software updates in these systems due to the lack of high-speed connectivity and interfaces (**Figure 5.4**).

Figure 5.4 SDVs with high-performance systems but with limited connectivity.

The platform approach will bring a significant advantage in reusing the technology and components across different vehicles. When different vehicle models are managed for different regions, reusability will reduce engineering costs and the cost of components inside the vehicle. Customization required for various users and various regulatory frameworks can be managed through software with a good platform approach and modularly designed vehicles. This is one of the significant advantages of SDVs. This will drastically reduce the cost of the vehicle, which otherwise would fall on the customers.

5.3.
Evolution of Software Architecture and Application

Earlier in this chapter, we discussed separating the software into a distinct layer in the architecture of SDCAVs. Detaching the software from being tightly coupled to hardware components is the power of SDVs. This allows further customization and updates of various features in the vehicle. Modularity gained using this approach will facilitate rapid innovation and deployment through software updates [5.2, 5.3].

It is important to understand the various software architectures being followed and how they influence the development of SDVs. We will explore the evolution of software architecture from being distributed across various hardware components that provided various functionalities to a centralized approach with software taking the lead role and being hardware-agnostic. Understanding how they ultimately bring down the cost of SDVs to their customers is also important.

5.3.1.
Monolithic Architecture

Monolithic software architectures were widely used and followed in the earlier days, where the concept of modularity was not considered, and thus this architecture had a significant dependency on hardware [5.4, 5.5]. This architecture incorporates all application components into a single unified codebase operating over hardware.

The codebase is tightly bound to the hardware. The data access layer, logic for the function, and the interface to the end user or any external communication are all merged into a single software component, as shown in **Figure 5.5**. Even if multiple functions need to be part of the architecture, they are merged into a single codebase using the hardware.

Figure 5.5 Monolithic software architecture.

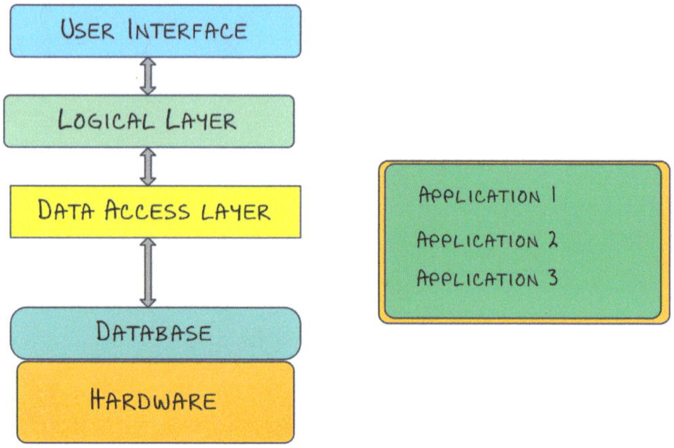

Monolithic architecture is a simple architecture that became popular for its lower complexity and higher performance because of its tight coupling with the hardware layer. There is no need for an additional mechanism to allocate hardware usage or manage data flow in these architectures. Monolithic architectures are still used for less complex hardware-dependent electronic components. The increased complexity of electronic components in vehicles, as well as complex ECUs and domain controllers, was found to be inefficient or unusable with this architecture, and we had to look for alternate architectures for the optimal use of the hardware and efficient functioning of the software. The drawback of monolithic architecture is that it is tightly coupled with each layer and hardware. Hence, changing a part of the function or customizing it becomes impossible, so the complete codebase needs to be changed.

5.3.2.
Modular-Monolithic Architecture

Modular-monolithic architecture is an advanced concept that allows software to be built into smaller modules rather than a single unit, unlike monolithic architecture. Here, flexibility is available as applications are built modularly, allowing for independent management and changing them as required. The database and the hardware layer below the software layer remain the same for all these independent modules.

The independent modules integrate their own data access layer, logical component, and external interfaces [5.5].

Figure 5.6 presents an overview of a modular-monolithic architecture showing different applications integrated as modules into a single hardware unit with a common database layer. Applications operate independently and can access the database, as they have a specific allocation for data in the database.

Figure 5.6 Modular-monolithic software architecture.

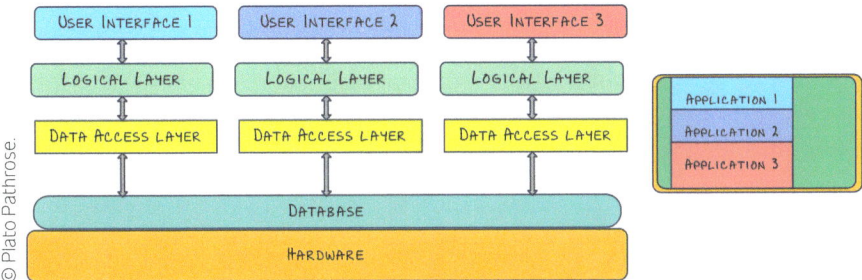

Modular-monolithic architecture is widely used in the automotive industry, where software can be deployed as a single unit. Still, internally, it integrates loosely coupled software modules that perform different functions. This architecture gives an advantage in managing the software as a single unit while keeping it modularly different in operation. Keeping the software modular allows the possibility to customize and scale the software and its functions up to a certain level. It is more robust and maintainable than other complex architectures.

Many of the safety components in vehicles utilize modular-monolithic architectures to be more reliable and robust. Even though there are many advantages to a modular-monolithic architecture, it is still limited in its ability to scale beyond certain limits. When we discuss how software-defined CPSs, such as domain controllers and zonal controllers with multiple microprocessors, are developed, we must also look at alternative architectures to address the scalability and data flow they should handle.

5.3.3.
Service-Oriented Architecture (SOA)

SDVs incorporate complex and scaled versions of CPSs. Monolithic and modular-monolithic systems may not fit well with scalable systems incorporating multiple processors and memory modules and performing multiple functions requiring high processing power. SOA focuses on modularity, where applications are built by combining standard, reusable, self-contained services that communicate with each other for information and data exchange and provide various functions to users.

Figure 5.7 presents an overview of a modular SOA, which has its own database for specific applications where services access and interact to provide functions. One of the significant advantages of an SOA is that it is loosely coupled to low-level hardware, and it can abstract low-level hardware details to make it easily maintainable. This brings flexibility in developing the software independently without being concerned about the hardware [5.5, 5.6].

Figure 5.7 Service-oriented software architecture.

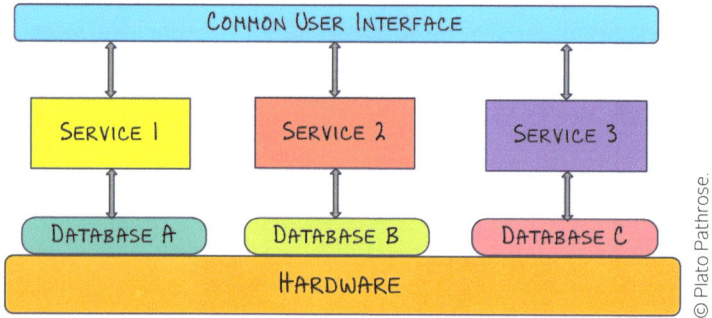

With vehicles getting more connectivity features for data exchange with the external world, SOA helps integrate external services that allow remote processing and exchange of data, such as data processing and exchange in a remote cloud data platform with an external infrastructure. It also facilitates the operation of OTA software updates, which act as the backbone of software updates and upgrades in SDCAVs.

AUTomotive Open System ARchitecture (AUTOSAR) and Adaptive AUTOSAR are some good examples of SOA.

Despite all these benefits and flexibilities, SOA raises concerns due to its complexity and manageability. Since it goes deep into the service level of software implementations, complexity becomes a significant concern, thereby bringing in chances of error during software development and deployment. Security is another concern that needs to be addressed. This is also due to the complexity with which communication happens in an SOA. Managing security during service calls from a server–client relationship becomes complex. Since many of these service calls may not be optimized during implementation and integration, this could affect the overall performance of the system and can bring in delays in data exchange. With all these complexities, governance and management are complicated when a system is fully implemented in an SOA.

5.3.4.
Microservices Architecture

Microservices architecture is an upgrade to SOA. Each application is further broken down into independent services, which introduces more standardized service components that contribute to building an application. Each of these services can perform specific tasks or have logic integrated within them, which can be used as standardized components across different applications [5.5]. This makes the architecture modular down to the most minor microservice levels that can be used in building various applications, performing different functions. This type of architecture generally brings in modularity and granularity for the software being developed for specific functions.

Unlike monolithic architecture, microservices are developed as separate codebases and they interact over simple interfaces. **Figure 5.8** presents an overview of microservices architecture and how communication happens when it is applied in building an application.

Figure 5.8 Microservice-based software architecture.

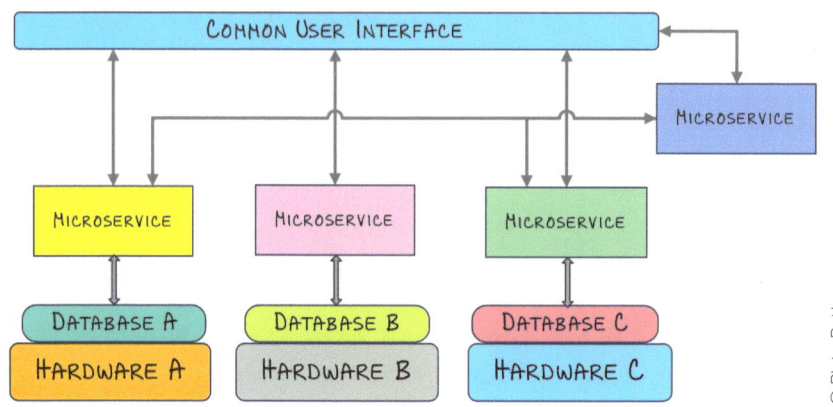

Microservices architecture is suitable for the latest technologies and scalability requirements we face today in SDVs, with numerous communication interfaces and data exchanges with the outside world. Since application interfaces are not bound to a specific function, rather configurable to use across various functions with various microservice calls, they also provide flexibility in resource allocation and database access as required. Microservice-based architecture is the result of the evolution of SOA that brings further flexibility and agility in building applications.

Microservices architecture is suitable for modern complex software development methods and achieving speed through reusability and standardization. Unlike traditional monolithic architecture, if there is a failure, it still affects only the specific application and not the entire function provided by the system. This helps identify and fix failures specific to services and applications without affecting the overall system operation. Microservices are widely used in data platforms such as cloud environments, where multiple applications operate at the same time without affecting each other. With these benefits, they are widely preferred for next-generation SDCAVs. Even though many manufacturers have started adopting them for specific vehicle systems, the complexity of managing microservices and the lack of standardization in the architecture used by various vehicle manufacturers prevent them from becoming widely used in the automotive industry.

5.3.5.
Lambda and Kappa Architecture

With technological advancements, we anticipate that SDCAVs will soon become moving data centers on wheels, collecting and processing a huge amount of data and sharing them with external stakeholders such as data platforms, infrastructure, etc. When vehicles start processing vast amounts of data, the software architectures discussed above might not become relevant or usable due to the delays they would bring to the overall data transfer and processing.

The lambda and kappa architectures are specific data transfer architectures that facilitate the movement of a significant amount of data with minimal delay. They are widely used in information technology to manage servers, data centers, and huge databases. **Figure 5.9** presents an overview of lambda and kappa architectures.

Figure 5.9 Lambda and kappa architectures.

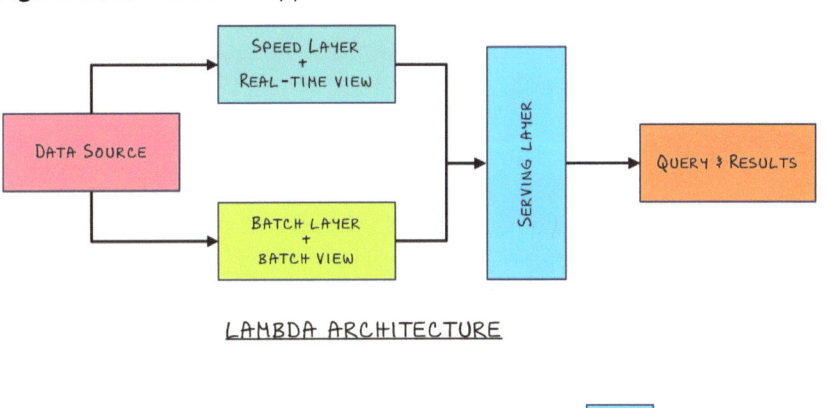

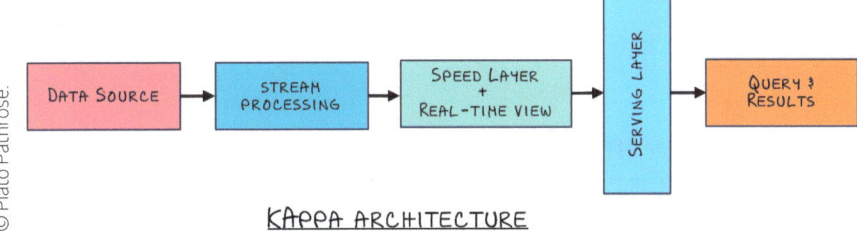

Lambda architecture is intended to balance the drawbacks that are usually faced around throughput, latency, and fault tolerances by leveraging parallel data paths for batch and speed data processing [5.7, 5.8]. The batch processing layer handles vast amounts of historical data, and the speed processing layer manages real-time data. Hence, real-time insights and updates are managed via the speed layer of the architecture. The serving layer acts as the bridge between these two layers, bringing the results together and enabling efficient querying and analysis of data as requested by the user or providing a live status.

Kappa architecture is a variation of lambda architecture. Here, instead of considering two layers for data flow, there is only one. It is a more streamlined and simplified approach to managing real-time data flow. The three layers in kappa architecture are the data ingestion layer, the stream processing layer, and the serving layer that addresses queries and results. When real-time data processing is needed, kappa architecture serves better than lambda architecture. However, lambda architecture allows the processing of a vast amount of data with historical insights, but at the cost of complexity of having a double layer and the cost to implement the same.

Many users might not consider that these two architectures can soon be viewed inside the vehicle. However, they may be familiar with these architecture types for various use cases, which require a lot of data collection and transmission for simulation and validation purposes. Autonomous driving functions are one of those functions that would generate vast amounts of data, which can be managed via these types of architecture for data transmission and analytics via remote data platforms and servers.

5.4.
An Overview of Vehicle Operating System

In the above sections, we have discussed the need for a platform and various software architectures that are used in the automotive industry for various purposes. Even though there are many advantages of different software architectures based on their applications, we can see a huge push among automotive players, from vehicle manufacturers to software stack suppliers, to start working on a dedicated vehicle OS. Why is it so?

Let us start with the purpose and the function of an OS before going through its application and relevance in SDSs and SDVs. The primary function of an embedded OS is to manage hardware and software resources in an embedded ecosystem. OSs support both real-time and non-real-time operations and services to enable multitasking, resource management, and reliability in their outcomes. We can have both a real-time OS (RTOS) and a general-purpose OS (GPOS) in embedded systems. RTOSs are fast and responsive within a certain time limit and are streamlined to operate in specific areas in a more specific way. They are also lightweight OSs, unlike GPOSs, whose goal is to provide results more quickly than other extended capabilities of OSs. The main functions of OSs can be summarized as follows:

1. Memory management.
2. Task management.
3. Inter-process communication.
4. Device management.
5. Managing timing and delays.
6. Network and communication support.
7. Security and reliability of communication and data.
8. Power management.
9. Support for different file systems.

With all these functionalities of OSs and the need for a platform that can be hardware-agnostic and application-agnostic in the development of SDSs, OSs become an important area of focus in SDSs.

The basic building block of an SDV is a software-defined CPS. A CPS can contain one or more microprocessors and microcontrollers with the same or a different set of OSs running on them to manage hardware and software resources and various other components that support its operations. OSs usually come from different vendors, and huge engineering efforts are required for their deployment and integration with various software and hardware components and optimization. Building an OS as part of a platform makes its deployment among various components easy and reduces engineering effort for integration and establishing the communication between different hardware and software elements associated with an SDS.

The primary reason why various vehicle manufacturers and software suppliers focus on establishing an independent vehicle OS is to reduce the heterogeneity in the lower layers of embedded systems, such as domain controllers or zonal controllers. Keeping the same OS among multiple microprocessors and even between different software-defined CPSs helps a lot in engineering efforts spent on integration and the possibility for reuse of various software elements.

The reusability and homogeneity introduced by using the same OS across major software-defined CPSs reduce the time taken for the deployment of most of the software applications and communication within and among these CPSs. It can be compared to a group of people communicating and sharing different tasks. Communication will be more effective and efficient when they all speak the same language and perform tasks in the same way. This avoids additional effort in organizing the data, the way they need to be presented to each person, and in communication, how the tasks need to be performed. The communication and speed of the execution of tasks will be more effective, reducing any additional effort and time.

The same logic of communication is used here, by adopting a common OS across various software-defined CPSs in the vehicle. This helps them own system operation and performance and reduces the time for integration between these systems in the vehicle. If they all speak the same language, it will be a significant help with the integration; otherwise, it would be difficult with different vendors, with changes that have to be managed over time for the complete vehicle lifecycle. An overview of vehicle OSs and how they influence the development of SDVs is shown in **Figure 5.10**.

Figure 5.10 Vehicle operating systems and their applications.

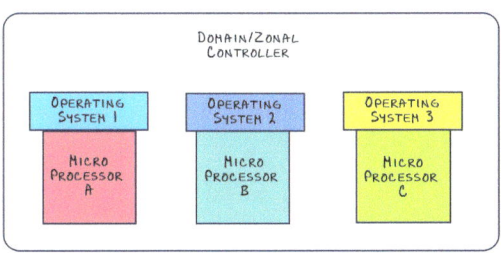

Each processor operates on different Operating systems within the same Domain/Zonal controller

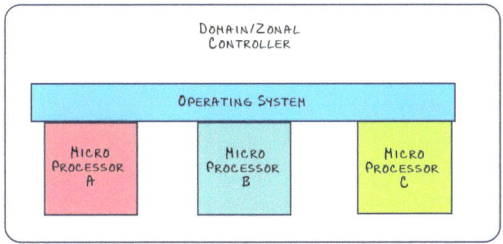

Same operating system used for all processors in a Domain/Zonal controller

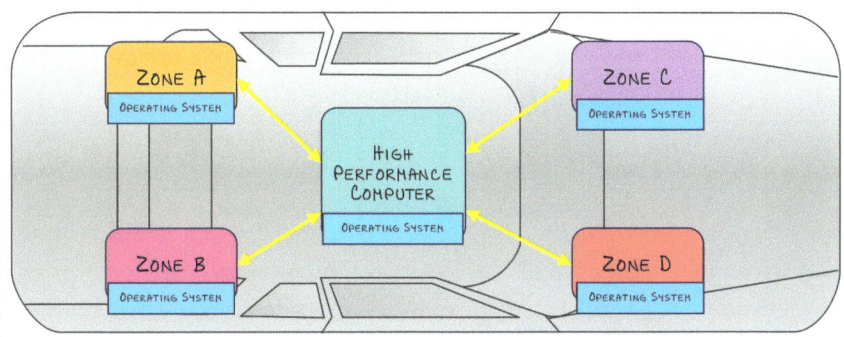

Common operating system across various Domain/Zonal Controllers in a vehicle.

5.5.
Open-Source Software in SDVs

SDCAVs have become too complex and expensive when considering various possibilities in which software can add value to the end user through frequent updates and upgrades to vehicle systems. Assume how much money vehicle manufacturers and their suppliers spend to build such systems to provide the user with a complete SDV with excellent functions. The challenge here is how such expensive stuff can be made available to their customers more economically, to be accessible to all users. With the complexity of these vehicles and their systems, building everything by a single stakeholder in the supply chain becomes a challenge.

Collaboration and using open-source software are solutions to address this challenge. They help build these complex systems and make them cost-effective and economically available to end users. Customers, being cost-sensitive, have significant concerns about cost when these software-rich complex systems are deployed in vehicles. Having an open-source software approach for SDVs will help save costs in different ways during the development, production, operation, and maintenance phases of the vehicle.

Open-source software components such as OSs, middleware, or even specific application components can be integrated into CPSs. This also supports quick iterative developments and innovation, as software and its source code are open to all and are available as standard repositories. Being open source would also gain a lot of support from the community and volunteers in addressing and resolving issues in software components quickly. This would not be the case if it were proprietary, in which case only a dedicated team from an organization with knowledge about software components would have to work on issues.

Customization based on different vehicles and the possibility of standardizing the interfaces and frameworks are other advantages of the adoption of open-source software in SDVs. Even though there are dependencies with specific legacy software components for hardware and interfaces in SDVs, using open-source software, things will change.

This will reduce the dependency on proprietary software and its license fees, thereby reducing the cost of development and in turn vehicle cost.

With all these possible benefits of open-source software components, some challenges must be addressed, which require solutions. Many players who opt for open-source software elements will question the maintenance and long-term support of community-developed software components. Since open-source software components are used across multiple systems, there is always a concern regarding software security and the overall security of systems utilizing these components. If there are dependencies with legacy software components in CPSs, which is the case today with the majority of vehicle manufacturers, integration with open-source software becomes a challenge. This may be because of undisclosed interfaces or software internals, or the noncompliance of the proprietary software to be integrated with an open-source component. In this case, the supplier of the proprietary software may have to be brought in to perform the engineering work and integration, which adds to the engineering costs in product development.

Even with all these challenges, the benefits of utilizing open-source software when building complex systems outweigh the disadvantages. The cost of product development will be significantly lower, facilitating the smooth transformation of SDSs and SDVs, which will become cost-effective to the users.

5.6.
Initiatives to Establish a Standard Architecture for SDVs

The complex nature of SDSs and SDVs can be addressed up to a certain extent through the standardization and utilization of open-source components for their development. This will also help new entrants to start their software-defined journey in a structured approach, with which they can develop and deploy systems with multiple functions, with quick integration into the market. In the automotive industry, a lot of standardization efforts are initiated by many players from different technology areas in SDVs. This section discusses some prominent standardization efforts that are driving the change in the ecosystem of SDVs and the automotive industry.

5.6.1.
Eclipse SDV Working Group

The Eclipse SDV working group is managed by the Eclipse Foundation, which focuses on establishing an open technology platform for developing and deploying SDSs and SDVs (https://sdv.eclipse.org/). The platform mostly focuses on software development, software orchestration management, and runtime for efficient data transmission. This also covers the services part in SDVs, where communication, data exchange, and management of multiple vehicles are taken care of using cloud platforms.

Eclipse Foundation uses a collaborative approach with the working group; they try to stick to open-source standards and frameworks, facilitating easy and quick usage of the framework in developing software components [5.9]. The framework aims to address both safety-critical and non-safety-critical elements in the vehicle, where software could add value to the vehicle and the end user. The working group is rich with participants from various tiers of the automotive supply chain, ranging from vehicle manufacturers to technology service and platform providers, all contributing to achieve this goal. The idea of the working group is to bring in expertise and know-how from all these tiers to establish the framework so that it would facilitate the development of SDSs and SDVs.

5.6.2.
SOAFEE (Scalable Open Architecture for Embedded Edge)

The SOAFEE Special Interest Group (SIG) is a broad industry-led collaboration of various organizations from the automotive and technology areas initiated by ARM Holdings plc. It focuses on building an open architecture for developing and deploying SDVs (https://www.soafee.io/). The key target of this SIG is to build an open and shared platform for all cloud-native applications in SDCAVs that use different hardware configurations. The aim of the SIG is to create a hardware-agnostic platform that operates by encompassing various services irrespective of the hardware layer.

SOAFEE, being an open community and a free-to-join SIG, makes it interesting to participate, contribute, and build the framework and use it. SOAFEE operates around four main pillars for deploying SDVs: (1) standardization, (2) new software architecture and methodologies, (3) industry collaboration, and (4) vehicle simulation [5.10]. Like any other working group, SOAFEE has a range of partners from different industry tiers, from vehicle manufacturers to technology providers. This facilitates the application of expertise from a broader community to framework development, which could facilitate standardized SDS and SDV development.

5.6.3.
Connected Vehicle Systems Alliance (COVESA)

The COVESA is an open and member-driven global technology alliance accelerating the full potential of connected vehicles and mobility ecosystem (https://covesa.global/). It focuses on establishing open standards and frameworks for connected mobility. Initially, it was established under the name of the GENIVI alliance. Many subgroups and projects have been part of the COVESA, ranging from data specification for signals to connected services [5.11].

The COVESA, which includes various industry players, focuses on SDVs and specific areas of the complete vehicle. Various projects executed as part of the COVESA exemplify the same. Vehicle data specifications, information service, electric charging, etc., are a few projects executed as part of the COVESA that help in standardization and have been applied in real-life projects.

5.6.4.
Hardware Abstraction Layer for Software-Defined Vehicles (HAL4SDV)

HAL4SDV is one of the prominent funded projects from the European Union (EU) that focuses on building an architecture for hardware abstraction to keep the software from being too dependent on the hardware layer (https://www.hal4sdv.eu/).

The main goal of HDL4SDV is to unify software interfaces and development methodologies. HAL4SDV enables abstracting software configuration from vehicle hardware for both safety-critical and non-safety-critical applications in future vehicles. The project includes most of the players from the automotive and technology sectors in the EU.

The project covers a holistic approach from development to deployment, including tooling. It is mainly based on software and connected services that utilize cloud-based development and deployment services in vehicles [5.12]. This EU-funded project is also expected to bring in specific processes and methodologies that can be followed in the future SDV era.

5.7. Summary

Architecture is one of the most important aspects of any product development. The importance of architecture has been discussed in this chapter at different scales. The classification of various vehicle systems and subsystems and how they can be logically mapped to different layers of an SDCAV has been discussed in detail. This opens the door to understanding architecture usage and its evolution on different layers, from traditional vehicles to latest SDVs. The evolution of the E/E architecture from distributed to centralized, such as zonal architectures, and why many vehicle manufacturers still depend on hybrid architecture and legacy elements, have been discussed.

As software has become a key element in SDVs, various software architectures and how they facilitate huge data transfers and connected services, with external data platforms, have been discussed in detail. An overview of various currently existing software architectures in various vehicle components, from traditional to future architectures that were adopted from the information technology industry for high-speed data transmission, has been presented.

Cost is one of the major concerns regarding the use of advanced technologies and connected services in the development and deployment of SDCAVs. The importance of open-source software and its utilization to address cost concerns and make the product cost-effective has been evaluated. An overview of vehicle OSs and their relevance in platform development, and associated benefits for SDVs, has been presented. Toward the end of the chapter, different initiatives that are popular in the industry to standardize the usage of open-source software and frameworks for the development and deployment of SDCAVs have been listed, which provided an overview of how the industry is driving the development of SDVs further in a collaborative manner.

With the great power that software gives to these new-generation vehicles and systems also comes great responsibility. With complex systems, enormous processing power, and connectivity to the external world, numerous concerns pop up around the safety and security of these systems and vehicles. These concerns must be addressed to preserve the safety and security of users, even though it is challenging. The next chapter discusses possible safety and security frameworks that can be applied, considering various use cases of SDVs.

References

5.1. Pathrose, P., *ADAS and Automated Driving: Systems Engineering* (Warrendale, PA: SAE International, 2024).

5.2. Lopez-Herrejon, R.E., Martinez, J., Assunção, W.K.G., Ziadi, T. et al., *Handbook of Re-Engineering Software Intensive Systems into Software Product Lines* (Cham: Springer, 2023).

5.3. QNX, "Software-Defined Vehicles," The Ultimate Guide, accessed March 25, 2025, https://blackberry.qnx.com/en/ultimate-guides/software-defined-vehicle.

5.4. Richards, M. and Ford, N., *Fundamentals of Software Architecture: An Engineering Approach* (Sebastopol, CA: O'Reilly Media, Inc, 2023).

5.5. Newman, S., *Monolith to Microservices: Evolutionary Patterns to Transform Your Monolith* (Beijing: O'Reilly, 2020).

5.6. Rumez, M., Grimm, D., Kriesten, R., and Sax, E., "An Overview of Automotive Service-Oriented Architectures and Implications for Security Countermeasures," *IEEE Access* 8 (2020): 221852-221870, doi:https://doi.org/10.1109/access.2020.3043070.

5.7. Sawyer, A., "Lambda and Kappa Architecture," Medium, May 31, 2024, accessed April 3, 2025, https://medium.com/@nydas/lambda-and-kappa-architecture-594c54d7c81f.

5.8. Marattha, P., "Kappa vs Lambda Architecture: A Detailed Comparison (2025)," Chaos Genius - Blog | Explore Databricks & Snowflake Tips, January 7, 2025, accessed April 3, 2025, https://www.chaosgenius.io/blog/kappa-vs-lambda-architecture/.

5.9. Software Defined Vehicle, "The Eclipse Foundation," accessed April 3, 2025, https://sdv.eclipse.org/.

5.10. Soafee.io, accessed April 3, 2025, https://www.soafee.io/about/charter.

5.11. COVESA.global, "COVESA Groups and Projects," January 22, 2025, accessed April 8, 2025, https://covesa.global/groups-projects/.

5.12. HAL4SDV, "About," accessed April 3, 2025, https://www.hal4sdv.eu/about.

Chapter 06

The Concept of Safety and Security

We have been discussing the concepts and development of complex systems and vehicles. SDCAVs are complex systems on wheels, like any time before. Their great complexity makes their development and deployment a great challenge. Safety and security are always a concern when complex systems are developed and deployed. Safety and security concerns should be taken seriously, with many fancy functionalities and connectivity playing a critical role in today's vehicles.

This chapter gives an overview of safety and security concepts that can be applied to SDCAVs. It aims to present a high-level idea and apply it in generating safety and security concepts for these vehicles. All regulations and applicable standards that are required for product development will be discussed in this chapter, which will assist in the development of SDVs and various methods involved in it. This chapter will explain why a different outlook is important for the safety and security framework of SDCAVs compared with traditional vehicles.

6.1.
General Overview of Safety in Vehicles

Vehicle safety is mainly influenced by the development process, as well as the application of specific standards and regulations. Certain regulations mandatorily require the use of safety standards, whereas some regulations recommend using safety standards as the best practice for vehicle development and deployment. There are differences between mandatory and recommended use of safety standards based on regulations specific to various geographical regions.

Due to the complexity and challenges in the deployment of SDCAVs, using safety standards should be considered mandatory rather than optional, and safety must be integrated into their development and deployment. The classical functional safety standard ISO 26262: Road Vehicles – Functional Safety and the standard ISO/PAS 21448: Road Vehicles – Safety of the Intended Functionality (SOTIF) [6.1, 6.2] are the two safety standards commonly used in product development. The former is a standard in use for all safety-critical applications in the vehicle, and the latter is specifically applicable to ADAS applications in the vehicle. However, since some goals can be achieved by applying the SOTIF standard concepts, this standard acts as a basis for any safety analysis for ADAS and automated driving functions these days. With complex systems and the use of AI elements in safety systems, another standard that has recently gained traction is ISO/PAS 8800, which focuses on the safety of AI components used in vehicles [6.3]. With respect to SDCAVs, there are a few additional standards to consider, such as ISO TS 5083 [6.4] and UL 4600 [6.5] (**Figure 6.1**).

The safety lifecycle in SDVs starts like any other vehicle program. The safety process is similar to the product development process, from concept to product deployment [6.6, 6.7]. This chapter will not present an in-depth analysis of the functional safety process as its intention is to provide an overview and a framework on how to perform various safety analyses and what parameters one must consider for the safety analysis of SDVs.

Figure 6.1 How to define safety for SDVs?

Illustrated by Ameya kiran.

The framework defined by the functional safety standard ISO 26262 acts as the main framework for vehicle safety. As per UNECE regulations, compliance with the safety of various vehicle systems is mandatory for vehicles deployed in the EU region as part of the type approval process. On the other hand, unlike the EU, the regulations in the US recommend using safety compliance as a best practice for various systems rather than making it mandatory. However, there are provisions to consider it as mandatory only when the state or federal authorities have explicitly stated it.

The safety process runs parallel to product development. All these standards use the V-model, which is widely used in automotive development. A product's development cycle processes have a companion process from the safety lifecycle. Depending on the project, it starts with item definition and hazard and risk analysis (HARA) at the concept phase and early development phase to generate specific safety goals and requirements, which will further translate to requirements for hardware and software in systems as technical safety requirements. The process continues until the development and evaluation of individual components and their management, and then, the product goes to production and is deployed in the market.

The safety lifecycle remains the same for SDVs. Still, specific areas should be considered for SDVs based on how an SDV is developed and managed, which is slightly different from the classical product development and deployment that we are familiar with.

6.2.
Challenges for Implementing Safety Framework in SDVs

Software is the critical component in SDSs and SDVs, so most vehicle manufacturers and their suppliers face a significant challenge when applying standard safety frameworks and processes to a software-rich system that is dynamic over its lifetime. Predicting the system's scalability and functions over the vehicle's lifecycle is challenging. This gets further complicated when vehicles have a high dependency on connectivity and a lot of data, and when information exchange is expected to happen in the vehicle over time, overwriting existing software and functions in various vehicle components.

Connectivity to various elements and software updates are common among SDCAVs [6.6]. However, the safety framework will not be foolproof if it does not cover the use cases and applications over the entire vehicle lifecycle. Clarifying item definitions as recommended by the functional safety standard and defining use cases that the vehicle and various vehicle systems are expected to address over their lifecycle are foundational needs for the safety of the vehicle.

One of the major challenges nowadays is that when a product is initially conceptualized and designed, the concept of software-defined approaches is not taken into consideration, and on the fly, the scope and use cases change, which brings in delays and lack of specific safety concepts addressing the newly amended areas and functions in the vehicle. The scalable and dynamic nature of SDSs in the vehicle significantly contributes to this challenge while establishing a safety framework. In many systems, the platform approach that was used in their development allows them to scale by additional integration of software in them, providing various functions. This brings in new features and interactions with external counterparts, such as data platforms, infrastructure, devices, etc.

With the introduction of new functions and approaches, various systems and the vehicle as a whole might not have been considered for specific new use cases, and with these changes, hazards could arise over time. With highly hybrid systems in place, changes and scalability might have been only planned and implemented for specific SDSs; simultaneously, other systems and devices must also be protected. One commonly used approach is to isolate these SDSs and their functionality from other safety-critical components in the vehicle so that the changes are only focused on these systems and other safety-critical systems in the vehicle function normally, with no unexpected risks and malfunctions. **Figure 6.2** presents an overview of various challenges one would face during the safety analysis of SDVs.

Figure 6.2 Overview of various challenges to safety analysis in SDVs.

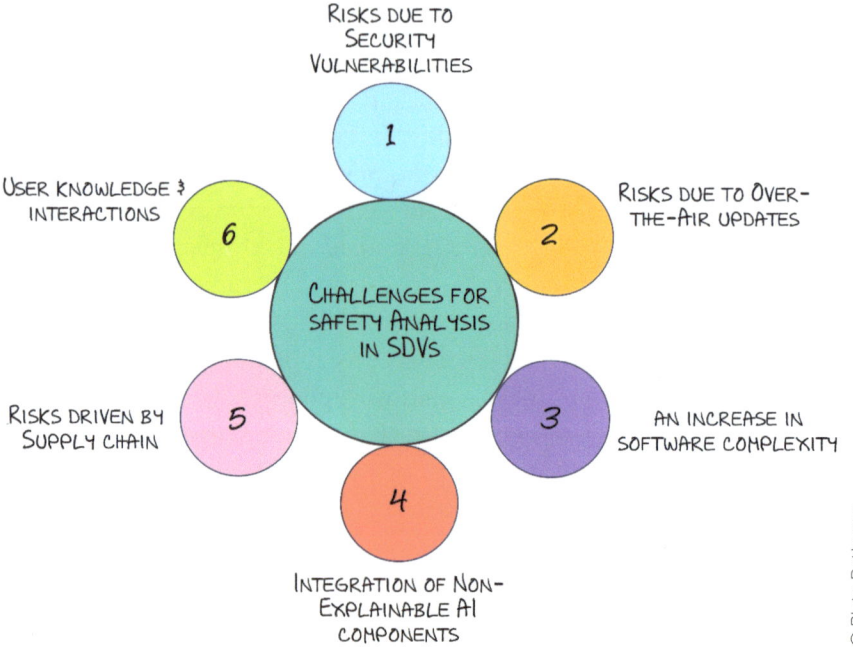

The challenges in the safety analysis and framework of these scalable SDSs and functions in the vehicle can be briefly summarized as follows:

1. Safety risks and challenges triggered by security vulnerabilities: With the dependency of SDCAVs on connectivity to external platforms and infrastructures, the area of interactions and the actors to be considered for analysis will increase significantly. This provides opportunities for malicious actors to utilize these possibilities to identify the vulnerabilities of the vehicle and its systems, thereby increasing the chances of cyberattacks in the ecosystem. If such cyberattacks target safety-critical systems in the vehicle, they can cause hazards and manipulate vehicle functionalities without the user being aware of it.

    Even though this is more of a cybersecurity concern for the vehicle, it overlaps with the safety of the vehicle. Hence, it is crucial to consider the safety and security of systems together when designing complex CPSs that are integrated into SDVs.

2. OTA updates during vehicle's operation phase: With connectivity as the backbone for SDVs, software updates and configuration changes

to vehicle functions are possible even after selling the vehicle to the customer. This brings in challenges and risks of new software upgrades and updates influencing safety-critical functionalities in the vehicle. New software components, configurations, and software updates could impact the safe operation of other functions in the vehicle if design considerations are not thoroughly analyzed and proper integration or regression tests are not performed before deployment.

The flexibility to perform OTA updates also encourages a frequent update and change mentality among vehicle manufacturers without considering the risks associated with it. Sometimes, software might not have undergone thorough verification and validation. The new functions planned might not have been analyzed for safety during the requirement generation. These gaps could be an area of weak safety considerations in these systems.

3. Increase in software complexity: As complex software-defined CPSs have become part of SDVs, the importance of software has grown significantly. The increased complexity also creates the need for complex architecture and the development of complex software components. With connectivity providing interfaces to external infrastructures and data platforms, the amount of software that establishes various services and functions has increased enormously compared with traditional vehicles. This increase in the lines of code and how codes are implemented on different architectures increases the chances of errors and bugs in the software during its development. These errors could be misconfiguration of software components and functions, wrong calibrations, unintended interactions, etc. The complexity of software is also one of the reasons for cyberattacks from a cybersecurity point of view.

4. Integration of AI components: ML and AI components have become standard elements in SDVs. With better performance and optimizations and having hardware with higher processing power, these fantastic software components have made their way into our daily lives through various means, and vehicles are one among them. Complexity and performance challenges of many complex operations have been addressed using AI and ML components. The challenge

with their application in SDVs is the lack of clarity in analyzing them in detail and on how an AI component operates during decision-making. If an AI component cannot be analyzed and is not predictable in its output, it can likely be a risk component for safety applications. A non-explainable AI component is a risk component if it is intended to operate in any safety functions in the vehicle. Things become complex when there is information exchange between external data platforms. It is always challenging in such situations to decide on the application of AI components in vehicles.

5. Supply chain and supplier ecosystem: Manufacturing and deployment of a vehicle requires multiple suppliers and products over various phases of its lifecycle, such as systems, software, tools, services, etc. The risks associated with suppliers and their deliverables are enormous for safety-critical systems and their functions. Good processes and a reliable supply chain ecosystem can manage and address these risks up to a certain extent. With complex systems, the major challenge is the integration of various components, whether software or hardware.

   Integration is the biggest challenge for today's vehicles. Unlike in the olden days, development methods have evolved from classical waterfall or V-model-based development to iterative agile methodologies. The changing requirements and sometimes even functions require adaptation in interfaces and software configurations for various vehicle systems. These might not reach all suppliers if there is a more significant supplier dependency on the product. Clear communication, safety analysis, stringent quality determinations, and transparent processes can help resolve these misalignments among suppliers.

6. Safety interactions with the user: When SDCAVs receive software updates and new functions after they were sold to the customer, there is always a risk that must be addressed regarding how various human interactions happen with the vehicle and its systems. These days, many autonomous driving functionalities are deployed in vehicles on a trial basis or as an FoD by the vehicle manufacturer; human interaction, risks associated with how the user can take over when there is a take-back request, how the user is going to be aware of it, various fallback mechanisms, etc., are all concerning if they are not known to the user.

Having precise fallback mechanisms, driver monitoring functionalities, and indications for the user for various actions becomes mandatory for the safe usage of vehicle functions. There are also regulatory implications when launching new safety features in the vehicle, which may vary depending on geographical locations. For example, if a function does not meet certain minimum quality and performance such as those evaluated according to General Safety Regulations (GSR), the user could be at risk in terms of their quality and performance while using the function. It would be risky if software-defined approaches allow an open path for deploying safety-critical functions without restrictions that require meeting minimum quality and performance.

6.3.
General Overview of Cybersecurity in SDVs

SDVs with connectivity and frequent software updates that provide various functionalities and opportunities to debug and remotely connect to external platforms over various lifecycle phases bring challenges and risks related to cybersecurity. In this section, we will discuss some of those challenges and how they can be managed in complex software-defined ecosystems. Having more connectivity options in the vehicle, such as Wi-Fi, 5G, Bluetooth, and wired networks like ethernet, CAN, LIN, etc., increases the probability of cyberattacks. There can be situations where some of the vulnerabilities are unidentified during the development and are unaddressed. Any malicious actors can use these vulnerabilities to gain access and attack vehicle systems and their functions, thereby affecting the smooth and reliable functioning of the vehicle.

What is cybersecurity in vehicles? It refers to protecting vehicles and their functions from cyberthreats to their E/E components [6.8, 6.9, 6.10]. Like the safety lifecycle, the cybersecurity lifecycle of the vehicle starts from its concept phase with an item definition through development and deployment [6.8]. Cybersecurity lifecycle is well defined in the standard ISO/SAE 21434, like the safety lifecycle as in ISO 26262. The challenges we discussed earlier in the safety lifecycle are analogous in the case of cybersecurity as well. Discussing various use cases and concepts covering all possible applications would be a good start to define cybersecurity requirements and their integration in the development of various components. There are regulations such as UNECE R155 and R156 [6.8, 6.11] that lay a fundamental framework for cybersecurity, and all vehicle manufacturers must meet them to keep data and information secure in the vehicle and prevent malicious attacks up to a certain extent.

Is that enough for modern vehicles? No! We must look beyond what the regulation says to address all possible challenges to the security of SDCAVs. Considering the complete lifecycle phases of the vehicle, based on use cases and applications where connectivity plays a significant role, we must keep a watch on the three main vehicle lifecycle phases and the complete organizational infrastructure and security processes to secure our vehicles. A general approach to how to bring security into a vehicle's development process can be briefly summarized as follows:

1. Control and secure attack surfaces: With connectivity and communication interfaces available for most of the systems in latest SDCAVs, the number of attack surfaces has been significantly increased. It can be an interface to the data platform, a third-party service, remote connectivity with the service center, etc. The interfaces that were in use during the production phase or those that are accessible at the service and maintenance phase of the vehicle could be a possible loophole for attackers if they are not well managed and closed after their use. The complexity of software in various systems in the vehicle and their interfaces, which are developed and integrated from various suppliers, could also increase attack surfaces. Open-source components used for many of these CPSs and services could be another attack surface. These open-source software components may not be specifically designed for the automotive industry and may be used in various other industries. Attackers can utilize the security glitches exposed by products in other industries to access and control automotive systems that use the same open-source components.

2. Securing the software supply chain: SDVs strongly depend on various suppliers and service providers for software and hardware components. These dependencies can increase the risks without transparency regarding software components and their interfaces

shared with the vehicle manufacturer. This is one of the challenges in vehicle programs when suppliers might not share knowledge on their software components and interfaces and claim them to be their intellectual properties (IPs). Any software component integrated into the system without understanding its interfaces and functionality is a security and safety risk. There could also be situations where the malware introduced from the supplier side to software components can impact the system's functionality. It could be via the supplier's development ecosystem or through their tools. Enforcing cybersecurity processes and security culture with all involved suppliers could help manage these risks. An open-ended supply chain with less control over the processes and security culture would be dangerous when such components are integrated into the vehicle.

3. Controlling remote access and configuration for various functions: Remote access and remote diagnostics have become essential functions of SDVs. Even though they provide enormous benefits, some risks can expose the vehicle to various types of attacks. Authenticating remote access and its functions is one of the processes that protects and manages risks from unauthorized access to the vehicle. However, if the authentication process is not strong enough, it could quickly become a weak link for any external unwanted authentication and gaining access to the vehicle's internals. Activation and deactivation processes of remote access and restricting the usage only when needed, along with time constraints, are essential requirements for remote access functions. This is one of the weak links in many vehicles that are vulnerable to cyberattacks. Either activation or deactivation processes might not be strong, or utilizing encryptions that can be quickly decrypted; these are some of the risks many systems suppliers and manufacturers face today.

4. Securing software updates and upgrades: OTA and FoD are two functions that help the vehicle to get new software for its systems and thereby avail updates and new functions. They are closely bound to connectivity to the vehicle manufacturer's database or server for remote connections for software downloads. Some of the cybersecurity challenges and prospective attacks that pose risks to these functionalities are man-in-the-middle attacks, invalid digital signatures in the downloaded software, and attacks around the rollback

functionality if faulty software needs to be rolled back to a previous version. Having a stringent process for OTA and FoD software updates and upgrades and establishing digital keys and signatures that can be validated for the downloaded software before flashing the software could address some of these risks.

5. Securing the AI components: SDCAVs use many neural network models to perform specific functions such as perception, decision-making, and driver monitoring. All these functions utilize data from various sensors, perceive the environment, and analyze the data for decision-making. The security concern here would be the possibility of tricking the neural network model with manipulated input data or training it with wrong data to influence its output. Many of these neural networks still have the drawback of having an architecture and decision-making capacities that are not explainable; this could be a loophole that attackers can utilize to manipulate and attack specific systems and vehicle functions. Having a stringent data policy for training and validation will help address these security risks.

6. Investing in and controlling organizational security processes: The vehicle manufacturer's infrastructure and facilities are critical to their vehicle's cybersecurity. A vehicle manufacturer needs many application interfaces in their production facility and service centers that connect with different supplier platforms, servers, etc. Moreover, many tools and interfaces are used for software flashing, diagnostics, and testing while the vehicle is in the production facility. The production facility must have a secure infrastructure with tools that are free from malware and bugs which could open an attack surface for attackers. Isolated infrastructure and processes that prevent the free flow of unscanned outside devices and software are key. At the same time, all external interfaces from the facility to the outside world should be well protected with firewalls and should have proper backups to manage data integrity. This helps address infrastructure-specific threats that can be later utilized by malicious agents and attackers to trigger attacks on vehicles. The mechanism to quickly roll back data from the backup when needed is vital if there is a cyber-attack. So, the infrastructure should focus on detecting and preventing the attack and also should provide a data retrieval mechanism for quickly retrieving the data from the database when needed.

7. Regulation and compliance for security and privacy: With cybersecurity for vehicles being of particular importance, specific regulations must be met for the vehicle to be market-ready for specific geographical regions. UNECE R155 and R156 are cybersecurity focused regulations that are applicable to the deployment of different vehicle types in the EU. The scope of these standards is to have certain cybersecurity solutions deployed within the vehicle and mandate the systems for managing its cybersecurity. Cybersecurity frameworks that have been required in the EU also cover organization-level management, reporting of cybersecurity incidents, and risk mitigation actions. This is being tracked and managed using a central database. Nowadays, with complex SDSs with high connectivity to external infrastructure and platforms, there is a huge risk if cybersecurity is only concentrated within the vehicle during its operational phase. Hence, the complete development cycle of the vehicle and its systems is being tracked and managed as part of the regulation.

8. Adopting best practices from other industries: CPSs with high connectivity capabilities have been used in the automotive industry for many years. However, unlike today, the need for security was not given much importance in the olden days. However, in the information technology domain, the knowledge and expertise around security have been in large-scale use for a long time. There are a lot of best practices from the information technology domain that can be used in the automotive industry, whether for the infrastructure or for the vehicle itself. Network isolation and trusted environments, secure boot, public vulnerability disclosure programs, regular pen testing, security audits, etc., have been some of the practices performed in information technology for a long time. These are not being used regularly in the automotive industry. However, adopting the standard ISO/SAE 21434 Road Vehicles – Cybersecurity Engineering [6.8] could be a good starting point, which guides development processes and incorporates security concepts as part of the design and deployment of the product. This is mandated by regulations in specific geographical regions, such as the EU, but still, many countries have not given due consideration to cybersecurity topics in the automotive industry.

6.4.
An Analysis of Interactions in SDCAVs

It is essential to know and plan the complete lifecycle of SDVs during the initial concept and design phase. This will bring significant advantages to product development and understanding of what use cases are planned to be addressed by the vehicle over its lifecycle. Considering the possible use cases and interactions with various vehicle stakeholders, **Figure 6.3** presents an overview of SDCAVs and their possible interactions over the operation phase and the service and maintenance phase of their lifecycle.

Figure 6.3 Overview of various interactions in SDCAVs.

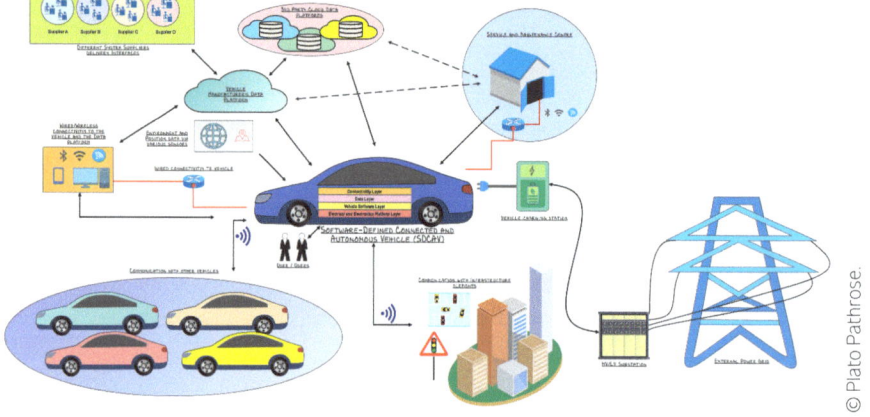

Here, we have considered an EV as an example. Considering an ICE or a hybrid vehicle as an SDCAV will not be quite different. Most vehicle manufacturers have initiated the transformation to SDVs by integrating SDSs into their existing vehicle platforms. This is attributable to the cost of completely redesigning their vehicle platform and the time it would consume to build and deploy an entirely new architecture in the market.

This is also driven by the lack of reliable SDSs and the infrastructure required to deploy them in vehicles in a cost-effective way. The supply chain mainly drives it as systems are still in the prototype and early production phases with most suppliers. Moreover, safety-critical systems with a different interface and software-defined approach require long testing and deployment phases, unlike less critical systems like infotainment or body control units.

In **Figure 6.3**, the operation phase of the vehicle is considered, where the vehicle interacts with the user, other vehicles, infrastructure, and data platforms for software updates and for subscribing to various services while using it. Being an EV, the interface with electric charging stations must also be considered across different locations. These interfaces act as a prospective attack surface if the vehicle is not well protected by its security interfaces. The services provided in the vehicle via third-party applications, subscription, and on-demand services all can act as additional risk elements while considering the possible threats that one must deal with if they are not well secured.

Depending on various use cases associated with various lifecycle phases of the vehicle, the number of stakeholders and interactions can increase. Here, for easy understanding, we have taken into account only the operation phase and the service and maintenance phase of the vehicle and its possible interactions. Each interaction will have an initiation process, a transmit/receive process, and a termination process. These processes are further detailed into smaller subprocesses and steps, such as protocol-specific message transmissions, key exchanges, etc. Depending on how successfully each of these processes are completed without any external interference determines the security of individual systems and the entire vehicle (**Figure 6.4**).

Figure 6.4 Vehicle communicating with the infrastructure.

Different connectivity features in the vehicle not only bring cybersecurity risks but also safety risks. Connectivity allows the free flow of information on both sides. As we discussed earlier, reliability and authenticity are very important so that they do not impact the functions of the vehicle. Once we know all the possible interfaces and interactions of an SDCAV over its complete lifecycle, it becomes easy to deploy a safety and security framework that covers these scenarios and interactions and helps keep the vehicle safe and secure. The next section will discuss a prospective model framework that can be used in SDCAVs, considering the interfaces identified during the operation phase and the service and maintenance phase of the vehicle lifecycle.

6.5.
An Overview of Safety and Security Concepts for SDCAVs

SDCAVs are complex systems with connectivity to various external systems and stakeholders over their lifetime. The complex software and hardware that perform safety-critical functions make it a system that needs to be thoroughly analyzed for safety during its development and deployment. Various connectivity options that come with these complex systems make things complex and require looking beyond the standard safety methodologies that we usually apply for safety-critical systems in a vehicle.

With the connectivity of the vehicle comes many attack planes. Security vulnerabilities can affect safety-relevant topics in these complex cyber-physical ecosystems. With many suppliers contributing to the vehicle manufacturer with their hardware, software, and tools associated with functionalities and connectivity in the vehicle, it becomes a complex scenario to analyze and evaluate possible safety and security risks and vulnerabilities during the development phase of the vehicle. Since the vehicle manufacturer is responsible for the safety and security of the vehicle, they should conduct a detailed analysis of the safety and security of SDCAVs, integrating all supplier systems.

Figure 6.5 shows a generic safety and security analysis framework, covering the possible use cases of SDCAVs. Here, the use cases are mostly in the operation phase and the service and maintenance phase of the vehicle. Whether use cases can be extended to the production phase of the vehicle needs to be checked, as vehicles are expected to perform specific functions, such as software driving additional capabilities in the vehicle, during the production phase. For this, the framework should also incorporate these use cases for specific production environments, and the associated stakeholders involved must be integrated for analysis.

Here, the proposed framework shows a two-zone analysis model that can be applied to SDCAVs. This model can be extended to autonomous mobility service providers providing services without direct human-driver involvement. It can also address various use cases related to remote and supervised autonomous driving, where the scope can also go beyond the usual road vehicles. The two-zone framework shown here is integrated with standards and guidelines usually adopted as part of safety-critical product development and deployment. Some of them are mandatorily recommended to be used by specific regulations for certain regions, such as the standards ISO 26262 and ISO 21434. However, some additional standards and guidelines come as a protective layer that would help analyze and safely deploy various functionalities in SDCAVs.

Figure 6.5 A generic framework for safety and security analysis of SDCAVs.

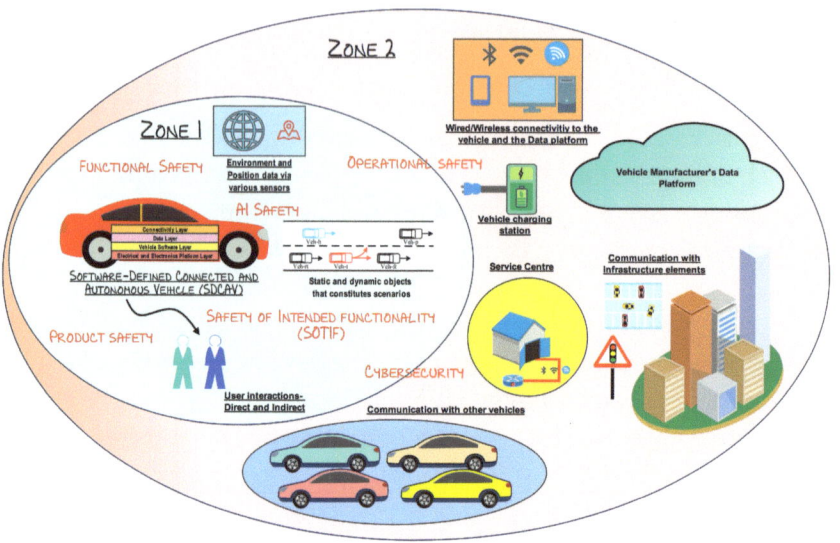

While analyzing various functions and use cases in SDVs, we can see a lot of overlap between the operation and the service and maintenance phases. Providing maintenance without delay is the strategy many vehicle manufacturers follow with the capabilities of these new vehicles. The two-zone framework starts with zone one, or the near zone, which is the immediate zone around the vehicle where most of the events occur that directly influence vehicle behavior. Zone two, or the far zone, involves a lot of interactions with various stakeholders based on connectivity in the vehicle. The zone-based splitting and classification of various stakeholders and infrastructure elements as part of these zones play a critical role in the analysis of the safety and security of SDVs.

The proposed framework uses the benefits of applying integrated concepts from various standards to analyze and design safety and security functionalities in SDCAVs. We could also observe here that specific standards overlap between the two zones. The zones are a virtual classification of various stakeholders. Positioning these stakeholders in zones will help better analyze and evaluate their influence on the vehicle at various applications. As discussed earlier, SDVs can be manually driven or fully autonomous, where software updates and upgrades bring in changes related to their functional performance and quality or even introduce new features. All these need to be addressed while considering a framework that covers both the safety and security of SDVs.

A general set of applicable safety and security standards for SDCAVs is summarized in **Figure 6.6**. However, this is only a basic set of standards. Many mandatory regulations specific to geographical regions must be met before deploying fully autonomous and connected vehicles; hence, the list must be adapted based on applicable use cases and deployment regions when considering this framework for the safety and security analysis of SDCAVs.

Figure 6.6 Overview of certain safety and security standards and regulations for SDCAVs.

In the following, we discuss some of the basic standards, regulations, and guidelines that are useful for analyzing SDCAVs.

1. ISO 26262 Road Vehicles – Functional Safety: This standard defines certain methods and guidelines to cover possible malfunctions in the vehicle's E/E parts that could cause possible hazards to the user and the environment. This is one of the classical safety standards that is mandatory to prove that vehicle components are safe in accordance with specific regulations in certain regions. The standard covers the complete lifecycle of the vehicle, from concept to production and deployment. It uses automotive safety integrity levels (ASIL) to classify risk and determine the required safety rigor. It follows a V-model development process like the classical development model used in the automotive industry, with steps for hazard analysis, system design, implementation, verification, and validation.

2. ISO/PAS 21448 Road Vehicles – SOTIF: Unlike the functional safety standard, SOTIF provides a general argument framework and guidance for measures to ensure the safety of the intended functionality without any unreasonable risk due to a hazard caused by functional insufficiencies. The standard addresses functional safety in systems where hazards may arise without component failure, particularly in ADAS and autonomous vehicles. Unlike ISO 26262, which focuses on hazards from hardware or software failures, SOTIF focuses on risks from performance limitations, unexpected environmental conditions, or misuse scenarios, even when the system operates as designed. ISO 21448 emphasizes ensuring that systems behave safely in all reasonably foreseeable situations. This includes understanding sensor limitations (e.g., camera blindness in fog), algorithm weaknesses (e.g., misclassification by AI), or edge-case scenarios not captured in traditional testing. This standard guides developers to identify potential hazards early, evaluate safety through testing, simulation, and analysis, and implement design improvements to minimize risks. It complements the functional safety standard to cover the vehicle's ADAS and autonomous driving functionalities.

3. **ISO/PAS 8800 Road Vehicles – Safety and Artificial Intelligence:** The ISO 8800 standard has become popular recently since its launch due to the dependency of AI components in complex CPSs in the vehicle. The safety of AI components has become an area of concern as they provide many benefits when used in complex systems. However, the lack of justification for some of the decision-making events and the output these AI components provide have become challenges to their usage in safety-critical applications. This standard aims to provide a framework for identifying, assessing, and mitigating risks specific to AI, including unpredictable behavior, data bias, lack of transparency, and learning system instability. Unlike traditional deterministic systems, AI components can evolve post-deployment during the operation phase of the vehicle, making safety and security assurance more complex. ISO 8800 promotes explainability, robustness, and human oversight throughout the lifecycle of AI components. It is designed to complement existing safety standards like ISO 26262 or ISO 21448 when AI components are integrated into safety-critical domains, helping ensure they operate safely, ethically, and reliably.

4. **ISO 21434 Road Vehicles – Cybersecurity Engineering:** The ISO 21434 standard defines processes to manage cybersecurity risks throughout the lifecycle of the vehicle. It starts from the concept phase to the EOL or the retirement phase of the vehicle. As vehicles become more connected and software-driven, such as SDCAVs, they are vulnerable to many cyberthreats that could affect the user's safety, privacy, and data integrity. This standard provides a framework for identifying threats, performing risk assessments, defining security goals, and implementing technical and organizational controls. It encompasses in-vehicle security systems and external interfaces that connect with the vehicle (e.g., infrastructure elements, data platforms, etc.). It ensures cybersecurity is embedded into automotive design, development, production, and post-production activities. The use of this standard has become mandatory for specific regulations for vehicle deployment.

5. **Operational safety:** Operational safety can be viewed as a broader application of the safety concepts in the above-discussed standards. It ensures the device or system remains safe during normal

operation, even when used under foreseeable misuse or varying environmental conditions. It includes broader concepts such as user interaction, environmental impact, fault tolerance, and behavioral safety (especially in dynamic systems like ADAS or automated driving), which have been partially covered as part of the standards like functional safety and SOTIF. Operational safety focuses on real-world performance, human factors, unintended consequences, and system behavior in edge cases on specific scenarios, even when no failure or malfunction is present. We could consider it an integrated approach of the above-mentioned safety standards and the standard IEC 61508 [6.12], including human factors and operational conditions of SDCAVs over their entire lifecycle and applications.

6. Product safety: Product safety is not considered a standard but is integrated as part of law in different countries. It ensures that a product does not harm users, property, or the environment during its entire lifecycle, including manufacturing, use, storage, and disposal. It addresses electrical, mechanical, chemical, thermal, and other hazardous risks. Regulations and standards help manufacturers design safe products and demonstrate compliance for each region. There are product safety regulations and laws for every country. One might ask why we have covered product safety. The answer is: it addresses the complete lifecycle and how the product must be managed when noncompliance occurs. Compliance with product safety includes risk assessments, testing, labeling, documentation, and recall procedures to ensure the product meets legal and safety obligations globally or for a specific region. Regulation 2023/988 is an example of a product safety regulation specific to the EU [6.13].

6.6.
Summary

The complexity of SDCAVs is increasing compared with what it was in the earlier days. Besides the complexity of software and systems in latest SDVs, adding connectivity to them becomes a significant challenge for vehicle manufacturers in building a safe and secure vehicle. This chapter discussed the importance of safety and security in new-generation SDCAVs. Even though some independent standards and regulations address various safety and cybersecurity aspects in vehicles, they must be considered from the initial phases of the product lifecycle.

This chapter presented an easy-to-use framework for analyzing the safety and security of SDCAVs. It incorporates all possible interfaces and interactions a connected and autonomous vehicle would have over its operation and service and maintenance phases. Many manufacturers have missed the complete picture of these interactions during the concept phase of the vehicle lifecycle. Understanding the safety and security of vehicles could also help various system suppliers who provide SDSs and software to vehicle manufacturers in analyzing and planning their development accordingly. With all the complexities and dependencies with various suppliers as part of the supply chain, this chapter also discussed various challenges around planning and implementing safety and cybersecurity measures in the vehicle. Different challenges have been discussed, and specific solutions have been proposed that could help gain insights into how the safety and cybersecurity lifecycle should be planned for SDCAVs. A set of fundamental standards, regulations, and guidelines have also been discussed, which can be used as an easy reference to kick-start the software-defined journey.

References

6.1. ISO, "ISO 26262:2018: Road Vehicles—Functional Safety," 2018.
6.2. ISO/PAS 21448, 1st Edition 2022 Road Vehicles – Safety of the Intended Functionality.
6.3. ISO/PAS 8800:2024 1st Edition, Road Vehicles – Safety and Artificial Intelligence.
6.4. ISO TS 5083:2025 Road Vehicles -Safety for Automated Driving Systems - Design Verification and Validation.
6.5. UL 4600 - Standard for Safety for the Evaluation of Autonomous Products, 2nd Edition, March 2022.

6.6. Pathrose, P., *ADAS and Automated Driving: A Practical Approach to Verification and Validation* (Warrendale, PA: SAE International, 2022).

6.7. ISO/IEC/IEEE 15288:2015 Systems and Software Engineering—System Life Cycle Processes.

6.8. ISO/SAE 21434:2021 Road Vehicles – Cybersecurity Engineering.

6.9. UNECE, "UN Regulation No. 155 - Cyber Security and Cyber Security Management System," accessed April 8, 2025, https://unece.org/transport/documents/2021/03/standards/un-regulation-no-155-cyber-security-and-cyber-security.

6.10. UNECE, "UN Regulation No. 156 - Software Update and Software Update Management System," accessed April 8, 2025, https://unece.org/transport/documents/2021/03/standards/un-regulation-no-156-software-update-and-software-update.

6.11. Pathrose, P., *ADAS and Automated Driving: Systems Engineering* (Warrendale, PA: SAE International, 2024).

6.12. IEC, "IEC 61508: 2010 Functional Safety of Electrical/Electronic/Programmable Electronic Safety-Related Systems."

6.13. EUR, "Regulation - 2023/988 - En - EUR-Lex," accessed April 12, 2025, https://eur-lex.europa.eu/eli/reg/2023/988/oj/eng.

Chapter 07

The Trend Toward Shift-Left and Shift-Right

The days of traditional mechanical and hardware-driven systems and vehicles have passed, and we are now in the generation of SDVs that give us fantastic experiences and latest cutting-edge technologies during our journey. A commonly heard phrase among the engineering community these days is "shift-left" in product development. What does that mean?

This chapter will shed light on shift-left and shift-right approaches that are becoming common in the development of SDVs. It has been used in not only SDVs, but any of those CPSs where software is a critical component. In addition, time to market has become a priority for manufacturers. In this chapter, we will discuss the benefits and challenges of shift-left and shift-right approaches. These approaches may be beneficial only for specific areas and may not be useful for others. Depending on the product, application, and ecosystem, they must be tailored. The methods and approaches that we will discuss in this chapter are generic, as there is no single way to approach the design and development of software-defined CPSs these days. It always depends on the project and product; furthermore, project limitations and constraints must be considered when adopting any of those approaches in practice.

7.1.
An Overview of Traditional Product Development and Deployment

Many of us are familiar with the classical product development cycle in the automotive industry, which tends to follow the classical V-model or an iterative V-model for product development. Most of the standards that we use as references also use the V-model. The requirements and design are expected to be mature enough earlier in the development cycle, and the whole development is based on the designs and architecture laid out in the initial phase of the development.

Traditional development methods follow a methodology where hardware, mechanical, and software elements of the product are tightly bound, and it is extremely challenging to think about separating them and building them independently at different paces. Even though we start with system design and development, there is always an interdependency between hardware and software for various functions, development, and testing for deployment. It requires more time to identify whether the developed function is of good quality and performs well enough to be used in the system. **Figure 7.1** presents an overview of the classical V-model for the product that has been traditionally followed.

Figure 7.1 The traditional product development approach.

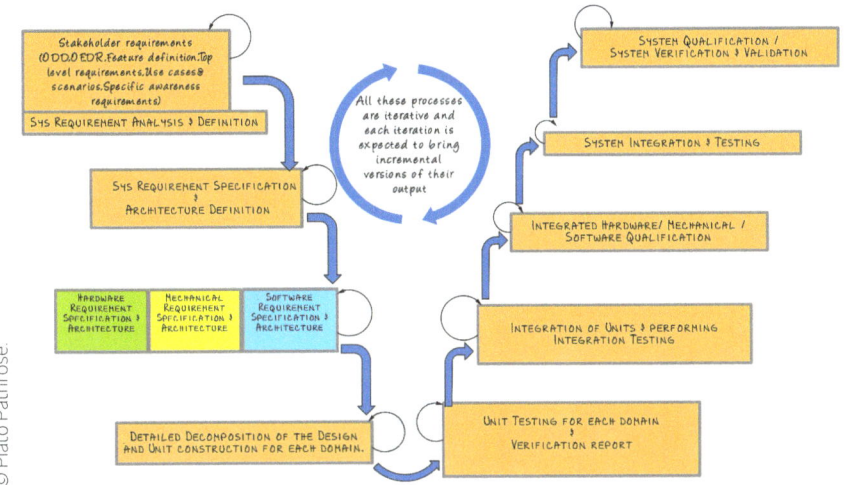

Traditional product development in the automotive industry was accelerated by adopting certain new development methods and processes from the information technology industry. Agile methodologies, which are widely used in software development and deployment in the information technology industry, found their way into the automotive industry. Thus, the importance of iterative development and the possibility of accelerated development and deployment of software in the automotive industry became popular. This improved the overall product development lifecycle and time to market for the product.

However, even with agile methodologies, the development of hardware and mechanical components was challenging to meet the speed and efficiency of software development. With the transformation of vehicles from a hardware and mechanical background toward being SDSs, these development cycles were not helping or bringing in the benefits of reduced development time or increased efficiency in overall product development and deployment. With the further adaptation of processes, methods, and technologies from information technology, specific hybrid approaches were tried for product development and deployment in the automotive industry.

The next sections of this chapter will discuss various methods and strategies adopted from information technology that have found their way into the automotive industry because of their benefits and advantages in developing and deploying software-dependent systems.
We will see how traditional V-model-based development, where there is a tight physical bonding between the software, hardware, and mechanical streams in the product, is being transformed when the automotive industry moves toward SDSs and SDVs.

7.2.
Digitalization and Changes from the Classical Product Development

Nowadays, there is a significant dependency on software in the automotive industry. This dependency is not just limited to product development or systems inside vehicles. It has a broader reach than we think. The development and testing ecosystem has changed from traditional to software-dependent ways. What drives these changes?

Since systems and vehicles have become more dependent on new technologies driven by software than ever before, the methods and processes widely used in information technology have easily found their applications in the automotive industry. This has started with the adaptation of processes in the development, deployment, and maintenance of software. The influence of the consumer industry, such as smartphones, has changed the way users prefer to use their vehicles. With all these changes in the ecosystem and technology, product development has also transformed from the olden times.

Digitalization is the primary enabler for this transformation in today's product development. It has changed how a product is developed, tested, and deployed. In the automotive industry, with the transformation to SDSs and SDVs merged with the digitalized ecosystem for product development, a lot of new elements are introduced into the product's lifecycle, such as connectivity to the internet, data platforms, digital development, testing, etc. About 15 years ago, one would have never thought about the various methods and approaches that we apply today.

The change in the environment, which is driven by the advancement of technology, including information technology, and the consumer industry, has influenced the automotive industry.

Looking at the development of today's automotive systems and vehicles, we can recognize a huge push to have a digital-first approach utilizing the concepts of digitalization. This approach means embedding digital technologies and experiences at the heart of how vehicles are conceived, designed, manufactured, and used. It is a shift from thinking of cars as purely mechanical machines to viewing them as software-defined, connected, and user-experience-driven platforms. Alternatively, this is also a mechanism in which product development can be started using tools for simulation and evaluation, even before building them in reality. The tightly bound hardware, software, and mechanical streams in product development seem to be detaching themselves from one another with the evolution of automotive systems these days. This is highly advantageous to the development of complex software-defined CPSs that are now the building blocks of SDVs.

The influence of digitalization and digital-first approach has made the development of software, hardware, and mechanical parts of the system easier. Many of the development and validation activities are possible for all these streams early in their development phase. The tools and the possibility to create a digital ecosystem to operate and model the system digitally help evaluate and identify the functional behavior and performance of each of the systems and components in the vehicle. This can be evaluated early in the development phase, even before any of these hardware, software, and mechanical components are built. Such a digital ecosystem, the digitalized version of systems, and even the complete vehicle in a digital environment can be called a digital twin [7.1, 7.2].

Digital twins are widely used in digital engineering, a multidisciplinary approach that uses digital technologies, tools, and processes to design, develop, produce, and deploy complex systems (**Figure 7.2**). It is one of the most popular and widely used approaches due to the influence of digitalization in product development and the advantages associated with it. The complete lifecycle of the vehicle and various systems can be digitally designed and evaluated for their correctness even before any real system development has been in place. Similarly, the ecosystem for development, testing, production, and usage environments can be digitally built to ensure that the product works perfectly in these environments. A digital twin is nothing but a digital replica of a physical system or SoS in a virtual environment [7.3, 7.4].

Figure 7.2 The concept of digital twins in the automotive industry.

Another recent change is that there are a lot of possibilities to scale up the engineering ecosystem. Earlier, any engineering development was closely tied to a particular technology, a location, or even an organization. In vehicle production, where hundreds of suppliers are involved, all suppliers are expected to deliver every piece of hardware and software to the vehicle manufacturer's facility during production. This was the same for an extended engineering development ecosystem involving many engineering service providers and software suppliers.

All suppliers were bound to operate and provide their deliverables to their customer location physically. This has changed with the utilization of data platforms. With advanced connectivity and flexibility to use enormous data storage and dedicated scalable processing capacities that can be remotely accessed across the world, the whole development ecosystem has changed.

Having such an infrastructure and the possibility to access tools by anyone, anywhere, and anytime opened enormous possibilities for the suppliers and other stakeholders involved in vehicle programs to develop and deploy software. Bringing connectivity into the vehicle made things much easier when it became possible for the vehicle to directly access data platforms, such as cloud-based data servers, to access software and download it as required. This allowed quick access to the latest software from any suppliers involved in vehicle development, which was made available to the vehicle user. Software can also be used to configure, update existing functions and system performance and bring in upgrades by introducing changes and new functions or providing additional capabilities to the user.

Briefly, the digitalization drive and availability of high-speed data exchange via connectivity played a significant role in transforming the automotive industry as we see it today. Software-defined approaches with various systems in the vehicle are a result of using the aforementioned technologies in the best possible way. Digitalization allowed not just the early development and deployment of software but also the adaptation of new technologies that can be used to improve the operational efficiency of various steps involved in software development and deployment. AI components and tools such as Generative AI (GenAI) are used in software development processes to generate data based on previously trained datasets or directly from various machine-based inputs for software design and development. All these facilitate the development of today's complex SDSs.

Since engineers have this much flexibility these days unlike before, it can be used to improve the whole process around how the product is developed and deployed. Digital engineering allows the design and development of not only software but also hardware and mechanical parts associated with the product. The possibility of having a virtual environment and virtual prototypes constituting the digital twin of hardware and software components helps check their performance and correctness even before they are developed. Many of the advanced tools with AI capabilities help generate a more reliable environment for performing these development and test activities.

Software's flexibility and non-dependency on hardware and mechanical components have helped develop and deploy it without considering whether hardware is available or not. Digital engineering methods such as utilizing digital twins and virtual models shorten development times, reduce the cost of engineering and costs associated with infrastructure and hardware, and improve overall product quality. All the architecture and platform adaptations and optimizations discussed in the previous chapters play a role in accommodating this change in the product development ecosystem and harnessing the benefits of quick development and deployment of software into various CPSs in the vehicle.

7.3.
The Concepts of Shift-Left and Shift-Right

The term "shift-left" is commonly used these days in discussions around SDVs. Why is this relevant? How is it used in the automotive industry? In this section, we will learn that and much more, explaining various concepts that are adopted in product development today regarding concepts of shift-left and shift-right.

The shift-left concept has nothing to do specifically with the automotive industry. It is a concept of performing specific activities or processes that generate valuable outcomes earlier in the product development lifecycle, which would otherwise have been done at a later stage. In the automotive industry, this concept adds much value and benefit for the vehicle manufacturer in getting certain activities executed earlier, which reduces the time to market and costs.

Based on the systems engineering lifecycle, vehicle development and deployment have multiple lifecycle phases, starting from the concept phase until the decommissioning or retirement phase. All these phases involve multiple processes that generate valuable outcomes, which are used in various other processes. The structure of any standard process can be defined using a SIPOC structure, as shown in **Figure 7.3**.

The input for the process is provided by one or more suppliers, which may be the outcome of other processes. In a process, various tasks will be executed based on input information and valuable data will be generated as output, which are supplied to specific customers. These customers could be the suppliers of other processes, providing inputs for executing other dependent processes.

Figure 7.3 SIPOC structure of a process.

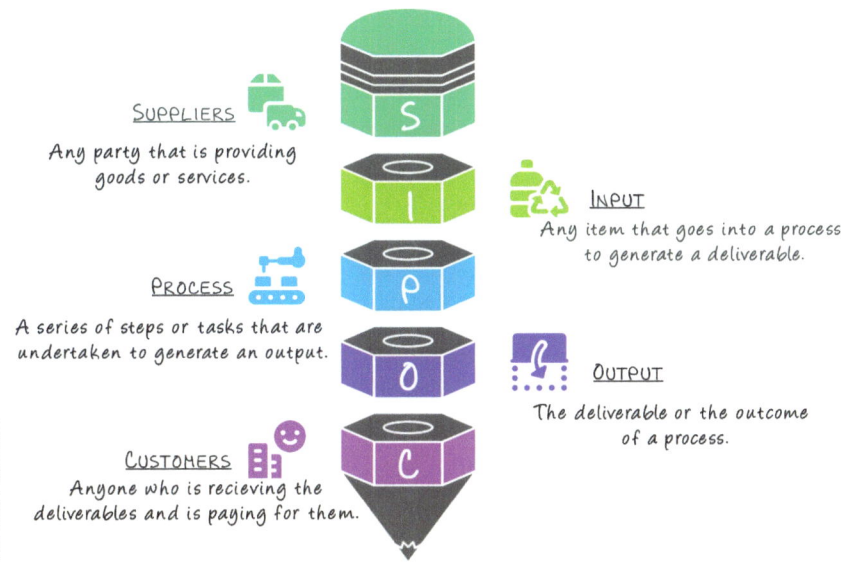

With the transformation in the development ecosystem, the adoption of agile methodologies in software development, and the benefits of approaching product development with digital engineering, the shift-left concept from classical software development is being reused in the automotive industry. It is applied in software development using agile methodologies, where testing is performed earlier, together with the development process, rather than waiting until the design and development of all software components are completed. This differs from the classical V-model widely used in the automotive industry for product development, where testing of software and systems is performed after software development is completed and integrated into hardware.

Adopting agile/DevOps methodologies, where software development happens iteratively, unlike the V-model, makes it easier to apply the shift-left concept. With this approach, the testing process that should normally be initiated after completing software development can be initiated earlier in the development phase of the lifecycle. This can be performed in parallel with the development of software components. This shift in the testing process not only results in early design changes but also engages the testing team to learn about the functional behavior for various test cases, based on which the design can be improved [7.5, 7.6].

As early testing is performed, the testing team actively participates in these earlier phases of development, thereby "shifting left" from the classical V-model process in software delivery. Performing earlier testing does not mean production tests can be skipped; rather, they will be more of a residual test, which takes less time than if it were following the process defined in accordance with the V-model. **Figure 7.4** provides an overview of applying the shift-left concept for testing. Test-driven development (TDD) and function-driven development (FDD) are some of the agile development methodologies that are being followed these days to harness the benefits of iterative development and the shift-left approach.

Figure 7.4 Shift-left concept for software testing.

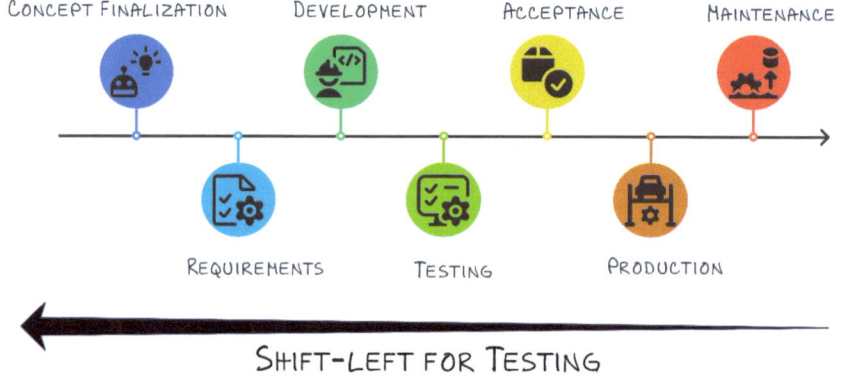

We have discussed how the shift-left concept is applied in software and system development during the development phase of the product lifecycle. Now, let us approach the shift-left concept on a macro-level. Specific processes can still be planned and executed in an earlier lifecycle phase. This means processes can be moved to an earlier phase if they add value to the product lifecycle. Digital engineering ecosystems and the usage of digital twins help us consider such a shift-left approach in the product development lifecycle. For example, design and testing of software, hardware, and mechanical elements and their functions can be performed even during the concept phase of the product lifecycle. Similarly, specific processes that are traditionally performed in the production phase can be moved ahead and performed at the development phase. **Figure 7.5** presents an overview of the macro-level approach of applying the shift-left and shift-right concepts in product development and deployment.

Figure 7.5 Shift-left and shift-right approaches.

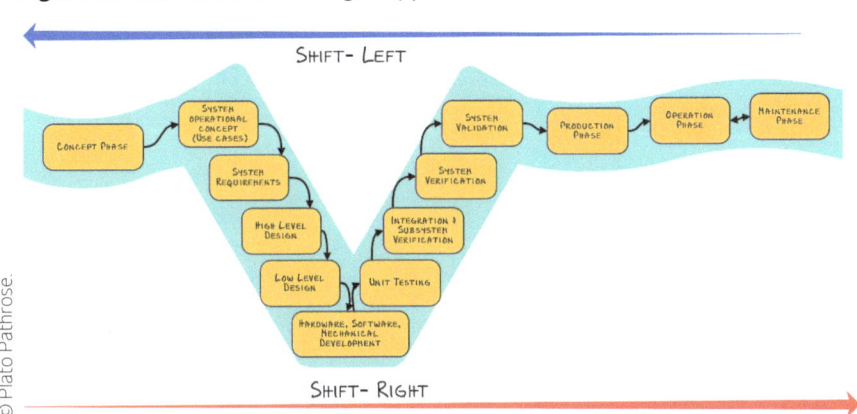

Similar to the shift-left concept, there is also a shift-right approach. It is not much discussed as it usually goes beyond software development to the production and operation phases of the product lifecycle on the macro-level. However, significant benefits can be achieved while applying this shift-right concept in the automotive industry. The shift-right concept is about executing specific processes that generate valuable outcomes later in the product development lifecycle. Usually, these processes would have been executed as part of an earlier lifecycle phase or performed earlier in the same phase of the product lifecycle [7.6].

For example, it is usual in the manufacturing industry to streamline production processes. This is one of the areas that is widely considered for identifying processes for the shift-right approach. Why is it always looked at the production phase? It is because setting up the infrastructure and executing specific processes in the production phase of the product lifecycle adds to the cost and time in production. Identifying these processes and optimizing them are classical process engineering aspects for saving infrastructure and engineering costs. SDVs have the advantage of having access to data platforms and digital ecosystems through embedded connectivity in them. Many of the processes from the late development or production phase can be moved to the vehicle's operation phase or service and maintenance phase, which can bring additional values and benefits.

The shift-left and shift-right concepts are not just randomly applied by selecting specific processes and steps in each product lifecycle phase. Instead, the decision to shift a process or task is made based on an in-depth analysis of the ecosystem and associated processes, their value generation, and the benefits of advancing or postponing the process in the same or different lifecycle phases (**Figure 7.6**). Many standard processes can be shifted based on knowledge from other industries. In the shift-left approach of the testing process, the benefits are well known in information technology for rapid software development and deployment, using agile methodologies. Process analysis and optimization to identify the shift-left and shift-right approaches are usually executed as part of process engineering by applying lean principles to reduce the wastage and bring value with any of the adaptations made. The DMAIC methodology in the six sigma approach is a commonly used process optimization methodology to identify and optimize specific processes [7.1, 7.7].

Figure 7.6 Deciding whether to consider shift-left or shift-right.

7.4.
Tools and Techniques for the Shift-Left Approach

It is always a question of what would help in adopting shift-left and shift-right concepts in the automotive industry. It is mainly the tools, techniques, and methods that are followed in an organization which drive efficiency by optimizing the product development and deployment processes. Some of the tools and techniques that can be applied in the shift-left approach are as follows:

1. **Model-based development:** The usage of models in the earlier design and development phase helps understand the requirements, evaluate the functional implementation, and test functional performance and behavior for specific use cases. Many tools available in the market allow an early design of the concept that addresses specific use cases with their functions and help implement models and test them. Model-based development and testing are techniques used in new-generation product development to understand the product

early and test design correctness and implementation feasibility. Using various tools, generating the software code directly from the models is also possible, thereby reducing development time and allowing to pretest the software for its functional correctness and implementation even during its model stage.

2. Digital twins: Digital twins are the result of the adoption of digital engineering in product development. They are virtual representations of physical systems and a dynamic environment with real-time data from various ecosystem counterparts. They consist of two parts, the digital twin prototype and the digital twin instance, which are integrated to exchange and utilize data from the digital twin environment.

With virtual prototypes and environments, developers and testers can virtually create an environment that matches the real world to evaluate system functionality and test various aspects of the product during development. Digital twins can be generated for individual components or the complete vehicle, and their interaction with the external environment and systems like traffic systems, environmental conditions, and infrastructure can be simulated. They are highly beneficial in developing complex CPSs such as ADAS and autonomous driving systems, where physical testing and evaluation for every possible scenario that the vehicle would experience during its lifecycle is impossible. The utilization of digital twins for development and testing will help shift-left many activities, even from the system-level testing, in the development phase.

3. Utilizing on-demand storage and computing: Smart infrastructure also facilitates shift-left approaches in the automotive industry. Remote cloud platforms facilitate on-demand storage and processing for the complete development and deployment phase of a product. On-demand storage and computing provide a scalable platform for performing various tasks in development and testing. This can range from a simple simulation of one of the software components to scaled and integrated testing of complete vehicle systems with virtual environments. Scalability also allows multiple instances to run simultaneously, improving development and testing times.

Deploying the complete toolchain for development and testing on cloud platforms allows easy integration and deployment of software components that are developed by different geographically distributed teams, which increases product development efficiency. Thus, it enables global collaboration and cost-effective and efficient resource usage in terms of licenses for tools and infrastructure that otherwise would have to be organized in each geographical location. Such a platform also allows secure and easier data management and quick iterations in software development and testing, shifting many activities to an earlier phase or process in the product development.

4. Applying AI as an enabler: The use of AI to perform certain tasks and processes is one of the greatest enablers in the automotive industry. It provides one of the best platforms to plan and apply shift-left approaches in complex product development. AI applications, integrated together with development and testing tools, make this proactive approach to testing and development possible. From requirement elicitation to model design and software development, large language models (LLMs) and GenAI tools have become enablers in overall software development. With respect to test case generation and implementation, utilizing GenAI reduces the testing time and improves resource utilization and coverage from an engineering perspective.

The early detection and analysis of defects using AI applications help identify and fix software bugs earlier in the development process, reducing effort, cost, and time. Using AI tools to support audits and process compliance checks, such as functional safety, ASPICE, cybersecurity, etc., reduces development and compliance struggles to a great extent, which would otherwise only have been possible to identify and fix later in the development process. From the project management perspective, AI-based tools help in project decision-making for quick changes and deployments, facilitating data-driven decision-making. We have tools that make our work easier on one side and also AI as an enabler on the other, which can be used in development and testing to improve and optimize work and performance. When these enablers can be integrated into different tools, the way work is done will change dramatically with improved timing, performance, and quality.

7.5.
Challenges Faced with the Shift-Left Approach in the Automotive Industry

The application of shift-left and shift-right approaches in the automotive industry brings great hope for transforming the whole development ecosystem and processes associated with SDVs as they move toward digitalization with advanced tools and infrastructure. Practically, these concepts work well with the software development stream. However, we must still address various challenges in applying them to vehicle manufacturing and deployment (**Figure 7.7**). Some of the common challenges and solutions that various stakeholders bring in to address them are discussed below.

Figure 7.7 Why cannot we move to a complete shift-left yet?

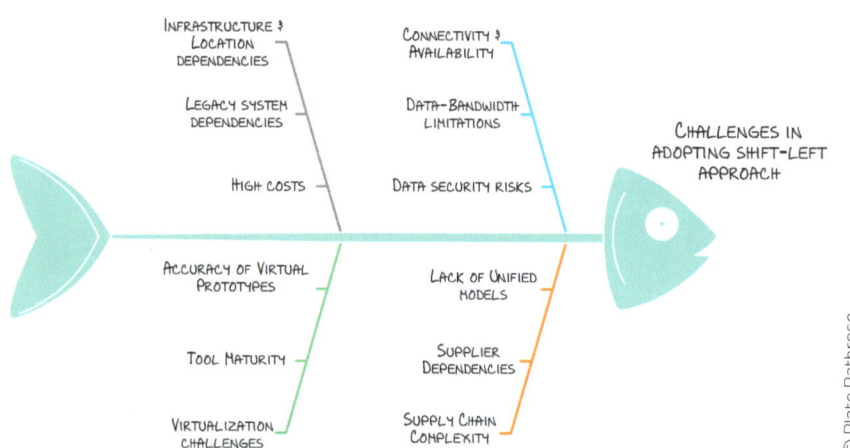

The infrastructure and processes involved in product development play an important role in adopting digital engineering concepts and applying the shift-left concept in testing and development. Many legacy vehicle manufacturers struggle with specific infrastructure dependencies that are part of their existing processes and ecosystem, where development and production occur. This prevents them from rapid transformation using digital environments and shift-left approaches.

In contrast, for emerging vehicle manufacturers, the lack of organizational dependencies with legacy systems allows them to adopt advanced tools and infrastructure in no time. However, the virtual environments being used usually come from various suppliers. The maturity of these tools and their capability in generating a real-world environment and replicating the behavior is a challenging question that often goes unanswered. Many players address this challenge by adopting a partial virtualization approach instead of migrating to full virtualization in their development and testing processes. The selection of which aspects need to be driven by virtualization and shift-left is to be made based on the value it would bring to each player in the product development supply chain.

Moving to digital engineering, deploying advanced tools, and adopting on-demand storage and processing capabilities brings in a lot of expense as part of the infrastructure in many organizations. Even though these approaches add a lot of value to the development and deployment of complex systems in the long term, there is always a cost associated with this migration. For legacy manufacturers, dependencies would be more significant as service agreements and contracts with the suppliers of existing tools and ecosystems might prevent them from migrating to a new framework or ecosystem. However, a gradual shift should be planned to control costs. Infrastructure costs are high, and they can be matched by understanding how they bring benefits in the long term. So, the decision should be made on gradual transformation, matching, and evaluating the best value it would give to the complete process and investments for the development and management of infrastructure.

The development and utilization of digital twins have significantly increased in recent years, as suppliers have started bringing up their virtual models to the market earlier than prototypes. Having a virtual environment to deploy software models and evaluate their performance and functions helps shift most of the testing earlier in the development phase. The possibility of building and deploying virtual ECUs and processors for software deployment reduces the development time. The concern is how much these virtual prototypes align with their real counterparts. The models of new processors that are not yet out in the market especially raise serious concerns about their behavior and the accuracy of their characteristics and performance compared to their real versions. The availability of accurate virtual prototypes and processors makes the application of the shift-left approach easier.

One of the key elements that drives digitalization and advanced engineering techniques is connectivity. However, there are challenges related to connectivity and data security with respect to on-demand infrastructures. Bandwidth utilization can be a limitation if an enormous amount of data need to be ingested or downloaded continuously. Such limitations add additional load to the network infrastructure and require a mechanism to control data utilization in on-demand servers. Having a reliable connection everywhere and every time is a challenge when we look at global collaboration as the main goal. Not having a strong connectivity infrastructure at all locations restricts the collaboration and speed.

The challenge related to data security and availability would also pop up if there is no adequate backup mechanism, even for a remote on-demand infrastructure, which adds costs and different service agreements. Management of such challenges requires additional skill sets that are not widely available in the automotive industry now. The lack of skill sets also prevents the adaptation of many digital engineering initiatives that could have been appropriate for many shift-left possibilities.

With all the above concerns, it is understood that the supply chain for vehicle manufacturing is very complex. There are many suppliers who have adopted advanced technologies; at the same time, there are many who have not. A complete shift-left using digital engineering methods and virtualization approaches for early development and testing requires infrastructure and virtual prototypes from various suppliers. This is one of the significant challenges, as suppliers might not be able to provide virtual models or the required information about their new components for usage in a virtual environment, as they may have strong dependencies on other suppliers or tools. Not having unified models and tools will add further complexity to the integration of models and even their large-scale adaptation in the automotive industry.

7.6.
Some of the Shift-Left and Shift-Right Approaches from the Industry

With the adoption of new technologies and methods in product development processes, we have seen how different organizations try to apply shift-left and shift-right approaches to their product development. Now, we will discuss some of the shift-left and shift-right approaches that have been applied to SDVs, utilizing the flexibility and power of software capabilities in various systems and vehicle connectivity. A summary of various applications is presented in **Figure 7.8**.

Figure 7.8 Various shift-left and shift-right approaches.

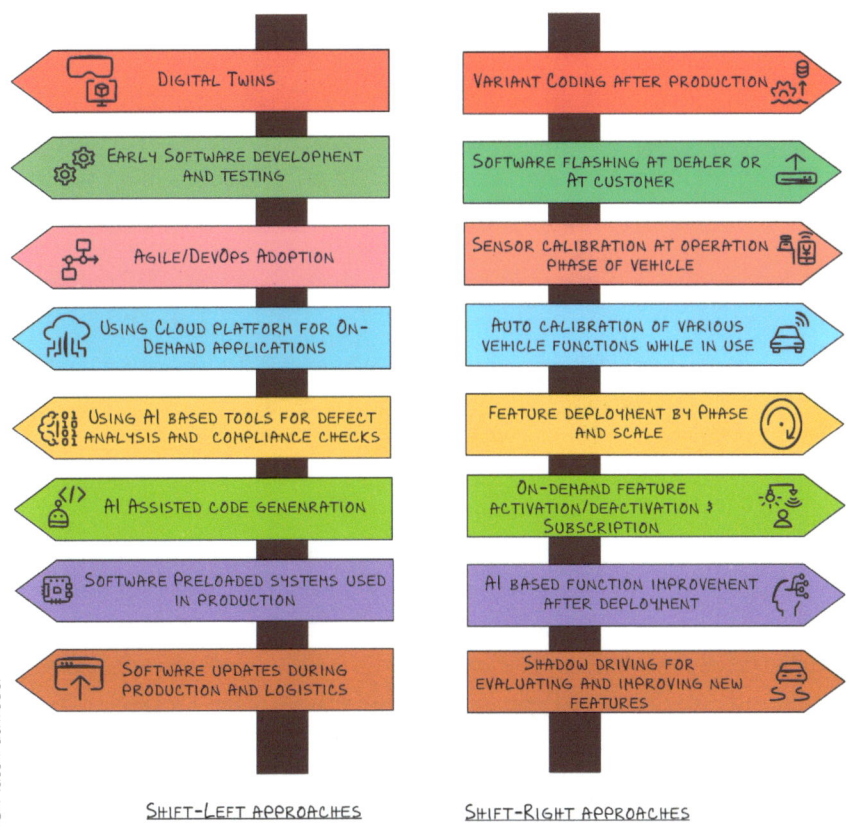

The change in the software and systems development ecosystem with digital engineering approaches, including digital twins and early virtual prototypes, has become a standard process in the development of many complex systems these days. Many vehicle manufacturers and system suppliers have started using iterative software development such as agile/DevOps methods to develop and test their software without waiting for real hardware in their early development phase. With the help of enablers like on-demand storage and processing and cloud environments with high-speed connectivity, many have started using tools and development processes with much better collaboration, speeding up the overall software development and testing processes. The introduction of AI-based tools for defect analysis and compliance checks in software development allows for the early identification and fixing of bugs and achieving compliance required for regulations, as well as safe and secure product development. Code generation and test cases have also been efficiently done using AI applications as enablers with various tools. All these play an essential role in the application of shift-left strategies in the development phase of an SDS.

Shift-left approaches are also widely applied on the macro-level, where processes or tasks are moved even to an earlier phase in the product lifecycle. One of them is delivering CPSs preloaded with specific configurations and software versions directly from the suppliers to the manufacturer's production facility. Many of you know that this has been a standard process done in production for quite a long time, and why is it special now? This is because even a preflashed system with a software version delivered by the supplier can be updated and upgraded with better software and new functions using connectivity and integration capabilities to remote data servers while the system is used in the production facility. This gives an extended time for software development and the possibility of having a better software version directly in the production location and during transport. Software updates would otherwise have only been possible during the service and maintenance phase of the vehicle. It is similar to any service subscription or activation, such as FoD and OTA software updates. All these shift processes from one lifecycle phase, such as the service and maintenance phase, to the operation phase.

Now, let us examine some of the shift-right approaches used in the vehicle manufacturing process. We have discussed various capabilities that connectivity and remote access to data servers can bring to SDVs. Specific tasks and processes can be shifted within the same or subsequent phases of the product development lifecycle. One of those applications is variant coding and location-specific features and services. Next-generation SDVs will be shipped as standard configuration vehicles to various locations. In a specific geographical region or a country, they will be flashed with specific software and required configurations for various functions applicable to that region, either in the dealer location or directly at the customer's place. This would be similar to configuring smartphones when we use them for the first time. Here, the process involved in the production phase is shifted to the operation phase or the service and maintenance phase.

Another shift-right approach is observed in the EoL calibration of sensors associated with ADAS and autonomous driving. Setting up an EoL calibration station in the production line is complex and expensive. Manufacturers have shifted this need from the production line to the operation phase of the vehicle, where sensors are configured to perform autocalibration before the vehicle is delivered to the customer or when the customer is using it for the first time. An adapted version is also used in service centers when repairs or replacements of sensors are needed during service. Vehicle manufacturers have also calibrated sensors in the vehicle using environmental targets on their path toward the storage area from the production line. These approaches also depend on the production facility and the steps followed by each vehicle manufacturer.

Shadow driving, on-demand activation, delayed deployment of vehicle features, etc., are some of the shift-right approaches widely used by many vehicle manufacturers. How and when they are deployed usually depends on vehicle manufacturers and their feature deployment strategies. Some deploy their vehicles with minimal features and add additional features after some time, saving on the development time and pushing those features directly from the development to the operation phase of the vehicle.

The deployment of AI components such as LLMs and personal assistants that improve with time and data is a technology enabler which brings improvements and experiences for SDV users. It is also one of the shift-right approaches in the automotive industry. In summary, shift-right and shift-left concepts are applied to facilitate the iterative development and deployment of products with improvements, along with optimizing the time and cost for developing and deploying these complex SDVs.

7.7. Summary

This chapter discussed shift-left and shift-right concepts in the development and deployment of SDVs. Adopting digital engineering techniques and an ecosystem driven by advanced technologies such as AI enables the automotive industry to implement more agile approaches in product development. The concepts from software development in the information technology industry were adopted into the automotive industry, with more importance given to software. This chapter explained how shift-left and shift-right concepts are applied on a micro- and macro-level within the same lifecycle phase and at different lifecycle phases of the vehicle.

The ecosystem and methods applied are the main pillars that help apply shift-left and shift-right concepts. Having tools and techniques drives the advancement in product development. Using industry enablers such as AI, various tools, and ecosystem-dependent techniques was discussed. Adopting shift-left and shift-right approaches will not help each vehicle manufacturer in the same way. All manufacturers have their own dependencies and challenges to consider while applying these concepts, which are mostly associated with their organizational processes and operational environments. In this chapter, various challenges and some solutions that were brought into use by different players in the industry have been discussed. A few examples of the application of shift-left and shift-right concepts have been discussed, considering the complete lifecycle of a vehicle. This will help any novice to understand the concept and recognize how these approaches can bring benefits to product development and deployment these days.

References

7.1. Pathrose, P., *ADAS and Automated Driving: Systems Engineering* (Warrendale, PA: SAE International, 2024).

7.2. Greengard, S., "Digital Twins Grow Up," Communications of the ACM, ACM News, August 6, 2019, accessed September, 29 2023, https://cacm.acm.org/news/238642-digital-twins-grow-up/fulltext.

7.3. SEBoK—Guide to the System Engineering Body of Knowledge, "Digital Engineering," accessed April 14, 2025, https://sebokwiki.org/wiki/Digital_Engineering.

7.4. Grieves, M. and Vickers, J., "Digital Twin: Mitigating Unpredictable, Undesirable Emergent Behavior in Complex Systems," in Kahlen, F.-J., Flumerfelt, S., and Alves, A. (eds.), *Trans-Disciplinary Perspectives on System Complexity* (Cham: Springer, 2016).

7.5. Red Hat, "Shift Left vs. Shift Right," n.d., accessed April 15, 2025, https://www.redhat.com/en/topics/devops/shift-left-vs-shift-right#types-of-shift-right-testing.

7.6. Gunja, S., "Shift Left vs Shift Right: A DevOps Mystery Solved," Dynatrace News, January 31, 2022, https://www.dynatrace.com/news/blog/what-is-shift-left-and-what-is-shift-right/.

7.7. Gitlow, H.S. et al., *Design for Six Sigma for Green Belts and Champions: Applications for Service Operations—Foundations, Tools, DMADV, Cases, and Certification* (Upper Saddle River, NJ: Pearson Prentice Hall, 2006).

Chapter 08

Transformation in Product Development

SDVs are the result of the changing technology paradigm and its ecosystem. Various industries have influenced the automotive industry in the past few decades. Information technology, industrial automation, consumer electronics, etc., are at the forefront. Adopting ideas and concepts from these industries also changed how vehicles are developed and manufactured these days. This chapter discusses some prominent changes in the automotive industry, with time to market being one of the critical goals. SDVs, with all those complexities, are changing how product development occurs in the industry. With software being a critical element in SDVs, the changes it brings to the classical automotive industry are extraordinary, which are reflected in all areas of the industry, from development and production to after-sales and the services offered. One could see the changes that new technologies and software have brought to the entire industry. Understanding these changes and adapting to the next technological evolution in the automotive industry is essential.

8.1.
The Shift from Hardware-First to Digital-First

Vehicles are a symbol of social status. In the olden days, people preferred to own vehicles with a masculine look, features, and excellent mechanical capabilities. This was decades ago; today, vehicles have become a sign of sustainability, style, comfort, and simplicity. It would be interesting to see how such a transformation was made possible. One of the critical elements that played a role in this transformation is software. Software and technologies have evolved, and they have enormous power now to bring about changes and make the vehicle dynamic, which was not thought of ever before in the automotive industry.

The automotive industry has evolved with the change in technology ecosystems and the influence of other industries. Earlier, one of the key advantages for any vehicle manufacturer was having knowledge on engine technology, which determined the value and power that a vehicle manufacturer had in the market. This has changed; organizations with software skills and those who are good at effectively utilizing enablers such as data platforms, AI, etc., in their product development and deployment hold the advantage. **Figure 8.1** presents an overview of how the powershift happened in the industry from the classical automotive era to the technology-driven era like today.

People's mindsets have changed due to the influence of other industries, such as information technology, industrial automation, telecommunication and consumer electronics. Nowadays, people do not want their vehicles to be static, which is boring and does not improve their engagement. Instead, they prefer dynamic vehicles with many visible changes that add more engagement and offer excitement and a better experience to them. This has resulted in the new generation of vehicles which we have today. Earlier, we had vehicles whose functions were determined by the hardware integrated into them. This has changed, and now, with software in the vehicle, it is possible to bring this dynamicity into the vehicle, including changes in functionalities, improved performance, quality, and changing UX.

Figure 8.1 The powershift graph.

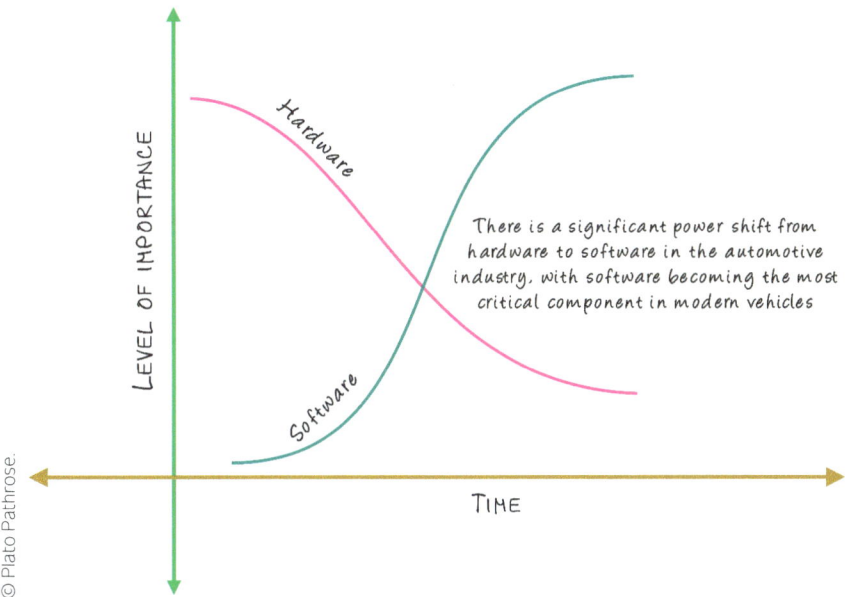

A traditional hardware-driven automotive industry has transformed into a software-driven one. The focus has now changed to software, such as how quickly it can be built, how are software changes managed, and what can software do more inside the vehicle than ever before. All these questions are answered by changing software development and deployment strategies, together with the introduction of software-driven CPSs used in modern SDVs. This is where the digital ecosystem has started flourishing in designing and developing complex software-defined CPSs. This has helped establish software as a separate stream without being a hardware-dependent one in building systems and vehicles.

Technological advancements and changing ecosystem with advanced tools and methodologies, which were not previously used in the automotive industry, have helped establish this software independence. Tools and various infrastructures, such as on-demand storage and processing, digital twins, etc., have provided a great platform to develop models of various systems and evaluate them for their performance and quality even before the actual product is developed, thereby having no dependency on hardware. Due to the time taken to develop, usually, hardware is one of the last components in a product development process for various systems in a vehicle, and this approach has dramatically solved that dependency by starting with the testing earlier, without even hardware being available [8.1, 8.2].

This change in product development methodologies using advanced tool ecosystems and enablers, such as cloud environments and AI components, speeds up the whole software development process. It gives the flexibility to build prototypes and test them even in virtual environments by creating the whole operating environment of the vehicle and various components in the digital world. It does not require software or hardware, but it gives quick feedback on how the software and the system will behave once it is developed and integrated into the vehicle. Adding any new features, functional optimizations, changes, and even bug fixes of existing systems can be performed earlier using this digital-first approach through establishing digital twins, such as virtual prototypes and virtual environments.

Most of the methodologies that are applied in quick and iterative software development can be applied in these digital models without even actually developing the real software. The data-driven design approach helps optimize the design and its implementation later in the hardware. This gives much flexibility to make design changes, which takes less effort compared to hardware-driven product development that was previously followed in the automotive industry.

Even though one would think that establishing a complete digital ecosystem for CPSs and even the full vehicle with all simulated environments is complex, it brings some significant advantages. Although the initial investment to establish such a digital ecosystem is slightly higher, the overall cost of development and reuse is very low in the long term. Customization and reusability allow digital prototypes and environments to be used longer than other methods that are followed in the industry, which can be more expensive. With early testing and improvements, utilizing the latest technologies and digital prototypes, design adaptation and implementation will be more data-driven than based on assumptions. This reduces the overall product development risk, especially when applying a new technology or deploying a new feature in the vehicle.

Figure 8.2 The digital-first shift in product development.

Improving the speed of data-based decision-making with a digital-first approach will automatically increase the speed of innovation without spending much time on unreliable concepts and approaches. This will improve the time spent on developing products using new technologies (**Figure 8.2**). Since a digital ecosystem with different enablers is scalable, its reusability and the possibility of applying it to multiple projects make it more economical and sustainable.

8.2.
Vehicle as a Development Platform and Data Center

The automotive industry has become a digital-first and software-driven industry. Software is the critical component that drives innovation, providing different experiences for vehicle users. What can software do inside the vehicle to make it dynamic and improve user engagement? Software performs various functions inside the vehicle as part of various critical components. With software, it is possible to update existing functions in the vehicle to improve their quality and performance and upgrade different systems inside the vehicle to enable or introduce new functionalities. Software can improve the way the user interacts with the vehicle. It can bring in additional services and capabilities and offer better personalization to the user. By connecting the vehicle with external infrastructures and platforms, software can bring many functionalities and services within the vehicle at any time. All these can be done with software inside the vehicle.

How can the vehicle manage different software components inside it to achieve all these benefits? That is a challenging question for many vehicle manufacturers. For better understanding, let us look at our smartphones. How do they operate now? We can add new applications, receive software updates for existing applications and OSs, enable new features, add new services, and personalize them for the look and the way they operate. The software drives all these changes within the smartphone. Here, the smartphone hardware acts as a platform providing optimal conditions for the software to operate and bring in these changes.

The same is happening in the automotive industry. There is a gradual transformation of the vehicle as a hardware platform as its foundation, and the software being agnostic to the hardware is bringing in all changes in that platform. A complex E/E platform that we have now, which includes various ECUs, domain controllers, and zonal controllers, is gradually moving to a consolidated hardware platform with fewer electronic systems integrated together. These hardware elements will be integrated like a smartphone, with the OS and low-level software just enough to make all hardware elements interact with each other and operate optimally with software deployed on them based on needs. Unlike in the earlier days, there is no need to separately manage each of these electronic components in the vehicle, as the whole control and management would be done with a high-speed processing computer, which acts as the gateway and command center of the vehicle. The evolution of vehicle E/E architecture to zone-based architecture highlights that consolidation, centralized control, and management will be the future [8.3].

The platform approach of the vehicle enables the vehicle to be further modular in structure, with each of the layers in SDVs to be further differentiated as separate streams in their development and deployment. Layered architecture, as discussed in Chapter 5, plays an important role in conceptualizing and adopting the vehicle as a platform with a significant amount of capabilities. With enablers such as AI components and the usage of cloud data platforms, vehicle connectivity allows seamless data flow from the vehicle to the outside environment and vice versa.

With many sensors integrated into the vehicle platform, the vehicle can capture all information from its surrounding environment. Depending on the sensors used, the sensor data are good enough to digitally recreate the whole environment around the vehicle. This can be processed within the vehicle or transmitted remotely to process and make inferences for specific actions that can be executed by the vehicle. Connectivity in the vehicle plays a major role in transferring huge amounts of data to the outside world for processing in a cloud environment. Each sensor generates a significant amount of data. When autonomous functions are integrated, and if these data need to be stored inside the vehicle or transmitted outside, it would be in the range of gigabytes of data per second. This is the approximate size of the data that vehicle computers are processing per second.

This gives us a different view of SDVs with complex sensor sets and autonomous driving functionalities. They generate enormous amounts of data. These data must be processed within the vehicle or transmitted outside for further processing and analysis to make data-driven decisions for performing vehicle functions. This is like a data center collecting and processing significant amounts of data. Next-generation vehicles have changed to a data center with a platform approach (**Figure 8.3**). The platform with hardware should have enough processing power and storage capacity if such amounts of data were to be stored and processed, and the data pipeline should be established that can be acted on by the software layer in the vehicle.

Figure 8.3 Vehicle as a data center.

Having remote connectivity to a data platform from the vehicle platform also adds benefits to SDVs. It allows functional deployments using software in a remote data platform rather than directly deploying in the vehicle systems. It also facilitates deploying additional services to the vehicle, fully operating from the cloud platform rather than transmitting it fully to the vehicle. Certain driving functionalities, such as low-speed functions, can be executed directly from the data platform with high-speed connectivity, thereby reducing dedicated hardware for this function. Many of such functions can be provided as a service-based or on-demand services rather than a direct purchase by the customer. This opens possibilities for new business models and delivery of multiple services inside the vehicle for different users.

8.3.
Enablers in Product Development and Testing

The complexity of developing and deploying SDVs is managed well with technology enablers. Next-generation vehicles are becoming more software-centric, with high-end hardware being deployed as part of the vehicle platform that supports scalability over many years. Software development and deployment have become a significant challenge. It would be extremely difficult and impossible if software development and deployment were to be followed using traditional methods. This is where these technology enablers come in handy in the new-generation software-defined journey. These enablers have found their way as part of the technology ecosystem and have gradually even found their way into regular operation of the vehicle, becoming an extended vehicle component operating even from outside the vehicle.

8.3.1.
AI as an Enabler

AI is one of the key enablers that has driven the evolution of technology in many industries. The automotive industry has recently started adopting AI components in product development and also integrated them as part of vehicle software to implement certain functions. The results are fantastic, opening a new realm of AI technology adoption. With the evolution and popularity of AI in various industries serving various use cases, it is essential to understand AI and apply it to product development. The evolution and commercialization of AI products using LLMs and GenAI tools have increased the influence of AI in different areas. Even though the ML-based object detection and natural language processing (NLP) solutions for interacting with the vehicle have already been popular in the automotive industry, new solutions with LLMs and GenAI are more like a delighter in product development. LLMs generate data based on text inputs they have been trained with, and GenAI creates new content, which can be text, image, code, or even videos. Nowadays, both are integrated with many tools that are widely used in product development and deployment processes in the automotive industry [8.4, 8.5].

LLMs and other GenAI solutions have many benefits and have been successfully applied in various other industries. They are now deeply integrated into the whole system and software development in the automotive industry. They reduce complexity, increase speed, and improve the product overall, with almost perfect solutions being delivered. The utilization of AI is widespread across various lifecycle phases of the product, mainly as a companion or a co-pilot in various phases. The development phase of a complex CPS starts with requirement elicitation and ends with verifying and validating the system to ensure it complies with the requirements and serves various use cases and needs for which it was designed. Each of these product development processes can be improved with the support of AI capabilities.

As we all know, the standard development phase of a CPS starts with requirement elicitation, and many lifecycle management tools are used from requirements to testing and traceability of those requirements. Building requirements from scratch is always challenging when a new technology is introduced or a product is built. Use cases defined for the product and various functions act as the foundation. This is where these AI-enabled tools can come in handy. They can help with an initial foundational set of requirements from the regulation, or specific standards could be listed with minimal time, unlike a human doing research and identifying each of those requirements, which can take hours of analysis. With these tools, a human can take a supervisory role to check the correctness and application of AI tools and their output.

Moving further toward requirement decomposition and allocation, these AI enablers could help generate the model and even generate the code in software or model development processes. They can even model and configure the system based on input requirements. This would be like a topping on the cake if error-free codes can be generated and evaluated for code correctness and compliance checks matching requirements. Many compliance checks must be considered in the automotive industry for the safety and security of applications, most of which are based on specific standards. AI tools could facilitate checking the compliance for the complete lifecycle, even during the development processes, and keep track of gaps and fix them while developing.

For testing, AI-enabled tools could help generate test cases, code for testing and performing automation, and scenarios based on test descriptions and even analyze defects and noncompliance resulting from test runs, indicating what needs to be corrected. Generating metrics and visual charts based on specific test runs supports project management in understanding and tracking activities and even organizing test executions and work distribution among the team. AI-enabled tools can benefit various project management activities involving different teams, such as generating project status, metrics, and charts for each stakeholder. The utilization of AI can be further extended to process optimization and automation and preparation of training materials for production and after-sales areas in vehicle production and deployment, which many organizations already consider these days.

With all these possibilities of using AI as an enabler in various steps in product development, it also results in a significant reduction in the time and effort that would have been spent if it had to be executed manually (**Figure 8.4**). This helps us do more things than we used to, unlike before, thereby reducing the time needed to build and deploy a product in the market. Efficiency improvement has been the focus here by integrating AI-enabled tools and systems into development and testing. With all these advantages, it is important that a human always needs to supervise and check the correctness of the data that have been generated and structured using these AI-integrated tools [8.6]. As LLMs and other GenAI tools are new AI systems being integrated into product development and deployment, there is always a chance of error. It depends on which model is being used and the accuracy with which it can generate data. Since many models are being deployed in the market, one cannot find something accurate and economical to meet all the needs. Thus, it is recommended that these AI tools be used sensibly as enablers and that one should not fall into the trap of blindly adapting them without supervision and correctness checks.

Figure 8.4 AI as an enabler in product development.

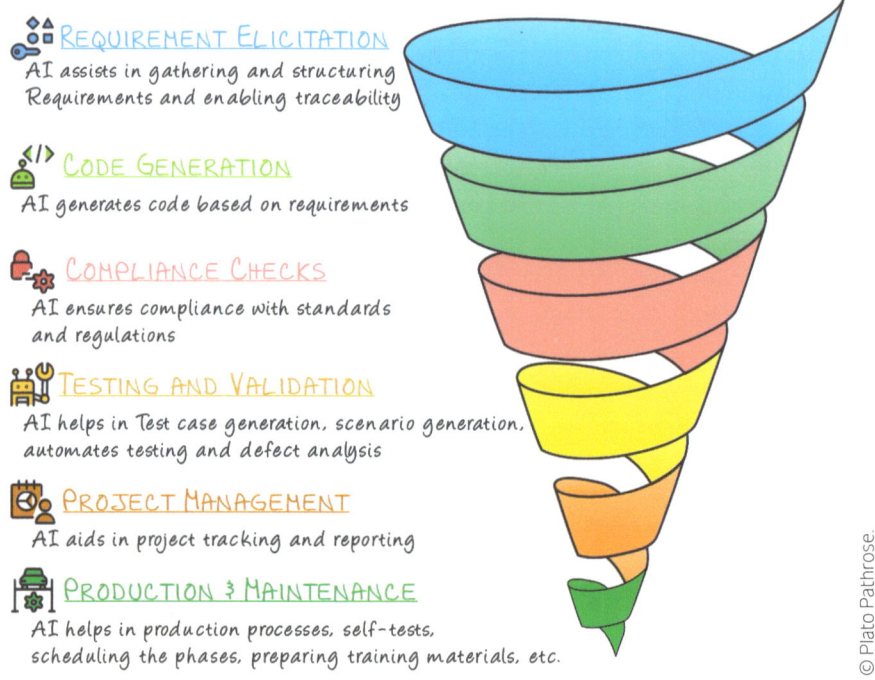

8.3.2.
Cloud-Based Data Platforms as an Enabler

Cloud-based data platforms are one of the infrastructure enablers that have found their use over almost all phases of product lifecycle. They are a scalable on-demand infrastructure that can provide unlimited storage and processing power, which is needed at every step in product development and deployment. It is not specific to the automotive industry and is widely used in information technology and industrial automation. With the changing ecosystem of vehicles being software dependent and generating vast amounts of data, cloud platforms have found their way into the automotive industry [8.7].

What would someone do when they can use unlimited storage and processing power? These cloud platforms provide on-demand storage and processing infrastructure. Their usage and use cases vary and are not limited to any process or lifecycle phase in product development.

Instead, with connectivity, it is possible to do many things using on-demand storage and processors. One of the main advantages of cloud-based platforms is that they promote collaboration and limit infrastructure drawbacks for geographically distributed teams and stakeholders. They reduce the costs of infrastructure by enabling simultaneous access to infrastructure with no limitations for different teams located in different countries or continents.

The usage of cloud-based platforms in the automotive industry includes remote platforms or servers where tools that support development and testing processes can be deployed. Even application and product lifecycle management tools can be deployed and accessed across multiple teams from multiple locations. When deploying other enablers on such a scalable platform, such as LLMs and other GenAI-enabled tools, these tools can perform better without constraints on the storage and processing power needed to provide a quick outcome.

Cloud-based platforms can even perform software development, integration, and automated test activities with the help of various tools. They can also act as data servers for delivering software packages on demand. For vehicles undergoing updates, it could be advantageous to use a cloud platform to host their software packages so that they can be accessed and delivered to a geographically distributed fleet through OTA updates with no downtime and provide users with FoD services, which otherwise would have been a challenge. Some vehicle manufacturers have even started using cloud platforms to host their operational software of the vehicle, where data are transmitted to the cloud platform from the vehicle, and processing is done at the cloud platform to send back the action points to the vehicle, so that the controls in the vehicle can act on them with no delay. With this approach, low-speed functions such as automated parking, valet parking, etc., are developed and are in their early deployment phase with few vehicle manufacturers. This approach is found to be successful with high-speed connectivity in the vehicle. There are many examples of using cloud platforms when coordinating with different suppliers and dependency packages from the concept phase to the service and maintenance phase of the product lifecycle [8.8].

A scalable infrastructure that provides unlimited storage and processing power brings great opportunities to optimize the development and deployment of complex CPSs. The use cases are not limited; one could decide how it can benefit them in their software-defined journey based on their needs and usage purpose. How will you use such a powerful infrastructure if you have it?

8.3.3.
Connectivity as an Enabler Bridge

We have been discussing SDVs and changing product development ecosystems. Many enablers, together with hardware and software streams, drive vehicles these days. Even though software remains the critical component in new-generation vehicles that are more dynamic, the bridge that connects the vehicle to the external world and makes all these things possible should not be forgotten—connectivity within and outside the vehicle.

Unlike traditional vehicles, we have discussed SDVs as data centers operating on software components. The vehicle's dynamic nature is only possible when it is managed with software updates and upgrades to engage and influence the user. Connectivity plays a critical role here. The E/E architecture of the vehicle lays the foundation for internal connectivity within the vehicle, with different harnesses and protocols for communication between various computers on the hardware layer. Information and data exchange within the vehicle while operating various processes to enable various functions helps provide different experiences to the user.

When there are dependencies with external components such as infrastructure, other vehicles, devices, etc., the vehicle needs high-speed connectivity to the outside world.

An SDV with a low-bandwidth connectivity is the same as a traditional vehicle with limited dynamicity. Software updates and size of data packages that can be exchanged between different stakeholders, such as infrastructure and data platforms, will be restricted by this limited data bandwidth. With high-speed connectivity and data bandwidth, dynamicity that can be brought inside the vehicle increases. Since the connectivity layer determines how dynamic the vehicle can be and how it can influence the overall vehicle operation, it can be considered as the bridge that connects the vehicle with all these enablers and other stakeholders in its operational ecosystem.

Connectivity remains a crucial component in transforming the vehicle or digitalizing it using various enablers for product development and deployment. It is impossible to use the capabilities of these enablers at their full scale when there is a bottleneck for data transmission. Considering all these aspects, the connectivity layer is critical in the software-defined journey of vehicles, which helps with the optimal usage of various enablers to improve efficiency and deploy the product.

8.4. Evolving Technology Adoption and Changing Skill Requirements

Evolution is gradual in any environment. The automotive industry also undergoes evolution in different areas, from engine technology to materials used for building vehicles. However, one component in the vehicle has undergone drastic evolution—the software inside the vehicle that performs various functions. This change drives the importance and need for software knowledge in many organizations, mainly for those that had previously operated with core hardware and mechanical knowledge.

Vehicles have also evolved with new technologies and the influence of digitalization and technologies from other industries. Nowadays, there are many EVs seen on roads compared to ICE vehicles. The movement toward sustainability and an environment-friendly approach drives this transformation. Replacing ICE vehicles with EVs and hybrid vehicles brings new technologies and the need for specific skill sets in the automotive industry. The importance of chemical engineering knowledge and the evolution of battery technology in different organizations are exciting. Impressively, battery technology too has software as the backbone for monitoring and managing its operation in the vehicle. Bringing in fascinating functionalities to the vehicle, such as autonomous driving and software-defined concepts that make the vehicle dynamic, opens up the need for new skill sets in the automotive industry.

Most of the transformational changes in the vehicle, such as its functionalities and how they are presented to the user, are managed through software. Hence, the automotive industry is gradually moving toward a software-driven industry like the information technology and automation industries. There will be more software engineers working in automotive organizations, unlike before, in the future. Integrating connectivity into the vehicle that allows high-speed information and data exchange within and outside the vehicle has opened a realm of possibilities that were not possible before. Along with these additional possibilities, the complexity of these vehicles has also increased. Since data transmission occurs between different entities within and outside the vehicle, it is also important to provide safety and security for the user who could be targets of specific cybersecurity attacks or hazards triggered by software, affecting vehicle operations.

With the connectivity layer of the vehicle, many enablers such as data platforms, on-demand infrastructures, etc., have come into the picture that were previously only used in the information technology or consumer industry. The automotive industry was never prepared to deal with such technologies in the past. This brought in many overlaps between the information technology and network engineering skill sets as part of the skills required for today's automotive industry. Unlimited storage and processing capacity allows many possibilities for both the development processes and the product after it is sold in the market.

AI, which was previously only part of the research community and applied mostly in the information technology area, is the enabler that has changed the world. It is not just the automotive industry that has benefited from this enabler. AI has found its way inside and outside the vehicle. These days, AI can be seen in everything that is operational and improves the efficiency of any tasks in our daily lives. Data scientists and AI engineers with software and data engineering skill sets are of high demand in the automotive industry. The effective use of AI, the value it brings to the overall product development process, and the improvement in efficiency are the driving factors for AI to become more popular and be adopted in the automotive industry.

With AI-based tools finding their way in each step of development and deployment, they have also brought in the need for new skill sets, such as prompt engineering and data analytics, as part of regular technology skills in the automotive industry. Prompt engineering is the process of structuring or crafting an instruction to produce the best possible output from AI systems. These AI systems could be LLMs or other GenAI-based tools that generate outputs that can be used to perform specific tasks in the development and deployment process of any product. Most of the tools used in the development and testing phases in the automotive industry are now integrated with AI systems to make them more efficient. So, it becomes obvious that in the future, prompt engineers will be expected to work together with system engineers in vehicle development, who act as the source for providing and making the tools operationally efficient with better prompts for accurate results that help system engineers.

The importance of hardware and mechanical engineering in the automotive industry is diminishing slightly and is being dominated by software and data engineering skills. Hardware and mechanical engineering may not be completely replaceable, but with the efficient use of AI-enabled tools, many of these tasks are made possible with software engineers with minimal support from hardware and mechanical engineers. Many mechanical topics are addressed these days, such as 3D printing and designing, using AI-enabled tools. All these being said, one should learn about the changing ecosystem and get themselves updated and upgraded with the skills in demand for tomorrow, especially learning to use these fantastic enablers that will change the world now or soon.

8.5. Summary

The change in the technological ecosystem and how it has transformed the automotive industry in building today's vehicles has been discussed in this chapter. The methodologies adopted from other industries, such as moving toward a digital-first approach for development and testing, has helped the industry improve the efficiency and the time it takes to bring a product to market. Giving importance to software and driving the concept of dynamic vehicle experience to users through software have opened new concepts and development approaches for the automotive industry. The vehicle as a platform concept is agnostic, and allowing software deployment with different services helps provide users with different vehicle capabilities when transforming to SDVs.

With the advancement of technology, we have seen some new technological enablers being adopted in the automotive industry.
We discussed in detail the usage of cloud platforms and AI-enabled tools and their application in establishing various vehicle functions. These enablers have also found their way in all the lifecycle phases of the vehicle. The bridge that connects the vehicle with all these enablers is the connectivity layer of the vehicle, which plays a crucial role in this transformational automotive era. With changing technologies in the automotive industry, it is essential for each one of us to understand and upgrade new skill sets that come in handy with next-generation vehicles. The immediate skill requirements would be about how to use these enablers efficiently in product development rather than how to build a product. As product development has been known to the industry for many decades, these enablers have changed the way it can be done, and the world is moving fast to adopt these new ways.

References

8.1. Slama, D., Nonnenmacher, A., and Irawan, T., *The Software-Defined Vehicle: A Digital-First Approach to Creating Next-Generation Experiences* (Sebastopol, CA: O'Reilly Media, Inc, 2023).

8.2. Pathrose, P., *ADAS and Automated Driving: Systems Engineering* (Warrendale, PA: SAE International, 2024).

8.3. Lionbridge, "The Car as a Platform. Driving into the Future," September 15, 2021, accessed April 17, 2025, https://www.lionbridge.com/blog/automotive/the-car-as-a-platform-for-technology-based-services/.

8.4. Deloitte, "Generative AI in Systems and Software Development: Breaking Barriers—Generative AI Leading the Charge," Deloitte Consulting GmbH, 2024.

8.5. Zao-Sanders, M., "How People Are Really Using Gen AI in 2025," *Harvard Business Review,* April 9, 2025, https://hbr.org/2025/04/how-people-are-really-using-gen-ai-in-2025.

8.6. Graydon, M.S. and Lehman, S.M., "Examining Proposed Uses of LLMs to Produce or Assess Assurance Arguments," NASA/TM–20250001849, NASA Langley Research Center, Hampton, VA, March 2025.

8.7. Kiefl, N., Wulle, F., Ackermann, C., and Holder, D., *Advances in Automotive Production Technology—Towards Software-Defined Manufacturing and Resilient Supply Chains: Stuttgart Conference on Automotive Production (SCAP2022)* (Cham: Springer, 2023).

8.8. Pathrose, P., *ADAS and Automated Driving: A Practical Approach to Verification and Validation* (Warrendale, PA: SAE International, 2022).

Chapter 09

Future Mobility and Transforming Business Models

What do you think about the future mobility ecosystem after understanding the change in the mobility industry and the influence of technologies and processes from other industries? It will be completely different from what we have today, as comfort and experience on a journey have become important and valuable in everyone's life, unlike before, and people are willing to consider that and pay for that. It is also driven by the changing lifestyle and how society has evolved. The change in mindset drives all these changes, and adopting these new comforts and experiences is an important element in one's life.

Let us look at the generation change. Technology evolution and its influence on the mobility sector can be visualized in three different generations. About 50 years ago, the older generation was more interested in reaching the destination, and the goal was to reach the destination by any means. It was driven mainly by the purpose of reaching the destination rather than the means of transport. That generation was not served with many options for alternative means of transport or comfort. The goal of reaching the destination was more important than the journey.

Things have evolved from there to a different level after 20 to 25 years in the automotive industry. Let us look at the generation about 20 to 25 years ago, who used vehicles for transportation. The living environments changed, and people started traveling more. The influence of globalization drove the need for business travel more than before. With the increase in options and means of transport, travel expenditures also decreased, and different possibilities were available for an individual to travel from one point to another or to reach their destination. With all these options, it was not just the destination that became important, but the journey too. Safety during the journey has become more of a delighter than ever before, and how safety can be provided was the main target for different mobility providers. Users preferred the journey that provided better safety. Thus, the importance of reaching the destination had grown from the goal of just reaching the destination to safely reaching the destination with no troubles. The evolution of vehicle safety systems started then and has become very popular among this generation.

Now, the present generation already has the mobility ecosystem that helps them reach their destination, unlike the first generation. They can now have a safe journey from the second generation, as vehicle manufacturers started offering various safety mechanisms in their vehicles. This eliminates the question of reaching the destination safely for a user. What the current generation is interested in is the experience they can have during their journey while reaching their destination. With different options and means of transport and technological advances available today, it is all about making a journey a wonderful experience with safety and security that makes a difference in everyone's life. The present generation is looking for that, and there are many travel options they can opt for in their journey. Future mobility providers will focus more on how they can be a differentiator in providing these journey experiences to their users.

9.1.
Change from Ownership to a Service

In earlier days, having a stylish, expensive vehicle was part of the status in society. People looked at the vehicle as an asset and the identity it could give to its owner. Unlike today, there were not many options for vehicles and modes of transportation available. One who owns a vehicle has more flexibility to travel and carry goods from one point to another. For longer journeys, people still prefer to use their vehicle at full or at least for part of it, even if they have to use other available modes of transportation like trains or buses. Until today, this is what is happening around us, but it is not very attractive for the new generation.

Generational change and the influence of new technology and processes from other industries are driving changes in people's perspectives these days. Many assume that owning a car prevents them from experiencing new features and experiences that other vehicles could offer. People do not think that buying an expensive car now is an asset, as its value depreciates over time. Urbanization and global ecosystem are driving the limit to owning individual vehicles, restricting them from finding space to park their vehicles or getting stuck in long traffic, etc. This hits back at users at their primary need for transportation. Having an expensive car with great functions does not change the situation, such as having no space to park the vehicle or getting stuck in traffic. Urbanization and global population increase have started affecting the primary goal of reaching the destination with a vehicle, as we can find more vehicles on the road these days. If reality becomes that the purpose of reaching the destination is not being met by owning a vehicle alone, then there is no purpose in owning it.

There needs to be a solution that allows people to reach their destination while having a safe and experience-rich journey. In these changing conditions, owning a vehicle might not serve the purpose everywhere. Different enablers, technologies, and the thought of optimized transportation have brought the concept of MaaS and on-demand mobility services to the transportation industry (**Figure 9.1**). MaaS integrates various transport and transport-related services into a single, comprehensive, and on-demand mobility service. It offers end users the added value of accessing mobility through a single application and payment channel [9.1, 9.2]. The integrated approach will help reduce complexity and provide a better experience for each multimodal transportation service, which will help the user reach the destination safely and with various comfort options provided by each service.

Figure 9.1 A multimodal and on-demand mobility ecosystem.

This multimodal transport infrastructure can involve various modes of transportation, from public transport, such as buses, trams, and trains, to micro-mobility services, such as e-scooters and e-bikes. On-demand mobility services such as manned or automated taxis and automated pods are the direct alternatives to owning a car in specific cases. The goal is to use the journey as part of the service offered by various on-demand transportation service providers. This would look similar to certain services we subscribe to or use on-demand on television or smartphones with specific applications. Both serve different purposes, as one provides options using multimodal mobility for reaching a destination, and the other uses a specific *ad hoc* approach with a single mode of transportation in providing the service when there is a need to reach the destination. Users can have different experiences offered by these services and select what they want to experience during their journey. With urbanization and technological introduction to mobility, both these and micro-mobility services are finding common ground among urban areas, and their adoption is becoming very popular among younger generations.

With all these available for today's generation, there is a massive transformation in people's thought process from owning a vehicle to using transportation as a service when needed. With different options and the availability of various players that provide mobility services in the market, integrating different modes of transportation is getting significant traction in today's developing and technologically advanced society. This is also becoming a mechanism to save costs on transportation rather than paying thousands of dollars in asset building by owning a vehicle, which depreciates drastically over time.

9.2. Personalization at All Levels

We have been discussing future mobility and various options that would be available for moving from one point to another, which are further enriched with different experiences during the journey. This is driven by software in today's vehicles. One of the key contributions of software to improving user engagement and experience inside the vehicle is providing personalization. For personal vehicles, various personalization features during the journey allow users to feel special, hence making them retain business and loyalty for the manufacturer. SDVs play a crucial role in utilizing software's power and specific enablers in bringing this personalization inside the vehicle.

Vehicle manufacturers select these vehicle functions that engage users more directly during the journey. The most selected functions for personalization inside a vehicle are infotainment systems, driver assistance systems, climate controls, seating, viewing units, etc. This is because the functions of these systems have a more direct influence on the user in the vehicle while driving or being a passenger, compared to other vehicle systems. Software can identify and utilize the preference for each user stored in the memory and make it available when the user is inside the vehicle, such as playing a favorite song, providing weather updates, traffic information, seating and mirror adjustments, etc.

With AI as an enabler and integrated into vehicle software, it could provide specific advanced personalization to users inside the vehicle. With the analysis of various data within the vehicle and of each user, it becomes possible to offer specific personalized solutions or custom functions based on the time and situation. This also increases user engagement with the vehicle and builds more trust in vehicle solutions. Anyone would love such personalization efforts when the vehicle recommends the user to take a break and organizes a halt during a long journey or when the vehicle automatically recommends specific destinations based on pattern analysis of the journey by a specific user. There are also AI-enabled functions, such as activating the refreshing mechanism in the vehicle periodically to keep the user engaged to avoid drowsiness during long drives. It will be interesting when the vehicle adjusts the ambient lighting conditions in the vehicle based on the user's mood, according to music, or on demand. How would you feel when the vehicle automatically recognizes your face and adjusts the infotainment system and services based on your profile from various cross-platform services or when it transfers data from other devices, which is currently extremely popular in this new era. Predictive personalization, the influence of providing various services such as payment services, parking-commerce, etc., within the vehicle on need, and integrating them with other vehicle functions are the future of personalization to provide a better usage experience for each vehicle user, irrespective of whether the user is a driver or passenger in the vehicle.

9.3.
Changing Business Models and Challenges

As the mobility ecosystem evolved, so did the various business models around it. It is not just selling a vehicle that has become important these days; many other factors drive businesses in this new mobility era. As discussed in earlier chapters, vehicle manufacturers are behind specific delighter functions in their vehicles that can provide a fantastic experience, safety, and security for their users. The advancement of digitalization and technology ecosystem drives various changes in the mobility ecosystem, which is no longer limited to the vehicle but involves many other environmental partners around the vehicle.

One of the most opted-for business models by various vehicle manufacturers is subscription-based services for their users. This can range from a simple subscription to a vehicle manufacturer's own or third-party service from the vehicle app store to an FoD that can be purchased for lifetime or for a specific duration as a subscription. Rather than providing all these services and features as a package and adding cost to the customer during vehicle purchase, certain subscription-based services have become a tool to reduce the cost of the vehicle, so that the customer does not face the cost burden when they buy the vehicle.

If we look at the way the industry is moving and delighters that various vehicle manufacturers are bringing to their vehicles using the power of software and connectivity in the vehicle, it is all about bringing a perception among users about the vehicle and various services they provide as better ones compared to others. With the evolution of the mobility ecosystem and the popularity of on-demand ride services and MaaS, the business is more focused on attracting their customers to use their platform and services for journeys. *"The demand will rise when service is perceived as a delicious food."* This classical quote on the service industry is very much valid in the mobility industry. Traveling from one point to another is not just a random use of the vehicle to meet the purpose, but rather an experience-rich journey with comfort, safety, and security.

The same has been put into practice with SDVs. The vehicle is offered as a service to subscribe to various third-party services like music, video streaming, etc., and commercial services that involve payments, integrating online e-commerce applications. Nowadays, the integration of certain AI-enabled personal assistant services and LLM models has become popular in vehicles to provide the experience of having a personal assistant who can help with anything as a companion during the journey. Subscribing to various services, purchasing upgrades, and upgrading the vehicle with new functions are some of the common business models these days.

Are these business models successful? It is yet to be evaluated from the vehicle manufacturer's perspective. A vehicle manufacturer's cost in providing these services becomes an overhead if they are a new entrant and do not have the required infrastructure and connectivity. For legacy manufacturers, upgrading the whole infrastructure brings additional cost and effort. Subscription-based services depend on connectivity in the vehicle, and this business model becomes successful only if it is operational for a longer duration due to the cost impact on the vehicle and the user who will pay for it. A study report from one of the consulting firms disclosed that the subscription-based business model is not successful for many vehicle manufacturers. Many vehicle manufacturers fail to establish a subscription-based model for long-term engagement with their customers [9.3]. It has a serious impact on the profitability of the vehicle manufacturer on their investment and the business they could acquire with that.

When a customer buys the vehicle, it usually comes with many subscription packages and connectivity. These are for a specific period, and the customer usually pays the cost as part of the vehicle purchase, which may not be made visible to them at the forefront. After the subscription period, only a few customers renew their subscriptions. In most cases, this is because customers are cost-sensitive and do not want to spend anything besides their vehicle cost, which they have paid initially while buying the vehicle (**Figure 9.2**). This breaks the profits of the subscription business model the vehicle manufacturer targets and the infrastructure they have already spent on. This is one of the significant challenges for almost all vehicle manufacturers on their journey toward SDVs and offering connected services. The race is on how these services can be offered to be perceived as *"delicious food"* to increase demand and continuous subscription. Many vehicle manufacturers fail to convey the value that these services bring to their customers, without which it would not be convincing for their users to continue subscribing to these services further.

Figure 9.2 Periodic payments and subscription-based services.

9.4. Data Monetization

Vehicles are becoming data centers moving on wheels. A significant amount of data is collected and generated within the vehicle. With all the infrastructure and facilities provided to the vehicle, to plan for the returns and profitability for which a vehicle manufacturer has invested, different ways are being analyzed to make the data from the vehicle a sellable product.

There have been different ways in which data are planned to be monetized (**Figure 9.3**). Many third-party vendors are interested in providing their services to vehicle users, customized to specific data and vehicle usage information. This starts with insurance companies that provide insurance coverage for vehicles with autonomous driving functionalities, where vehicle health status and driving conditions are used as input for tailoring their services specifically to usage-based coverage, thereby saving the cost to the customer through dynamic pricing. Commercial services utilize data and usage patterns to offer commercial services for vehicle users based on their profiles. This could be a tailored offering for a subscription for a particular duration or even for a journey. This could also be offering certain onboard shopping and payment experiences like parking, usage of e-commerce applications, and even ordering something from a nearby shop while in a foreign country with a different spoken language.

Figure 9.3 Various data monetization possibilities.

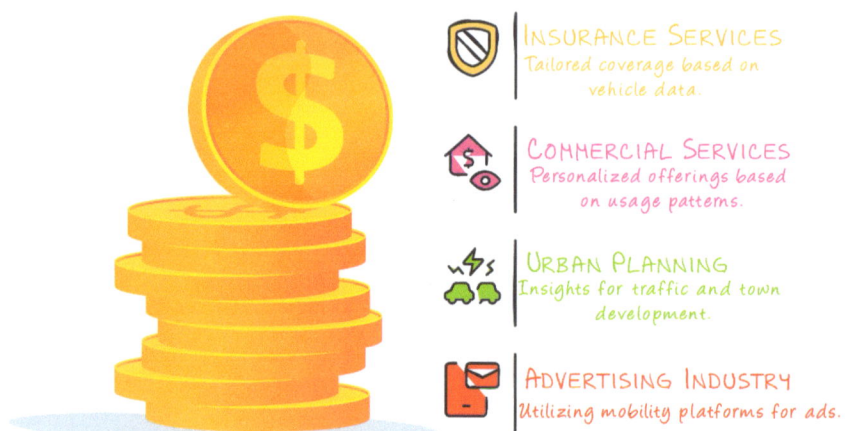

Environmental data captured by vehicle sensors could act as a good data repository for studying and analyzing traffic and town planning. Growing cities and town planners could utilize this data to help with their decisions around infrastructure planning based on the environment and usage data from various vehicles. Any specific service provider who wants to inform or deploy services in multiple vehicles could utilize the services offered by the vehicle manufacturer's backend data platforms to reach out to as many users in the vehicle. In the future, the advertising industry could use mobility-based data platforms as one of its key areas to reach out to its customers.

Since a lot of private information and data are involved, vehicle data, including environment data, and their disclosure are highly regulated in various countries. This must be respected, as data leaks could bring more harm than benefits if vehicle manufacturers and service providers do not give due consideration to data security and privacy topics.

9.5.
Which Business Model Should an OEM Prioritize?

The emergence of software-defined systems and vehicles opens up different possibilities for business with the technology and its byproduct from the vehicles, such as data. When we look around at different vehicle manufacturers, the subscription-based business model has become a foundational model among many. This business model may not be the best when looking at the long-term strategy and the drawbacks associated with what can be offered over time. Most of the revenue streams are subscription-based services that are offered to users. That relies heavily on connectivity. The connectivity does not come free of cost inside the vehicle; this has a dependency on the network providers in each region, and the cost of the network services indirectly falls on the users.

This creates a psychological detachment from the vehicle services over time, as it is making the user pay for the vehicle and the services associated with it for a longer duration. The decision of payment by a user can be completely based on the state of the user when such subscription services are about to be renewed.

For a vehicle manufacturer, this may not be a sustainable business model. When the only source of revenue is based on service subscription and when it does not bring in revenues, and there is no chance that the vehicle manufacturer can have any returns for the huge investments that they make in SDVs (**Figure 9.4**). This revenue generation challenge can be addressed by considering opportunities to generate revenue from various sources. The user can be one among them, but there can be more players who could benefit and may be willing to pay for certain information and data that they use to run their business from the vehicle and its users.

Figure 9.4 Which business model should an OEM focus on?

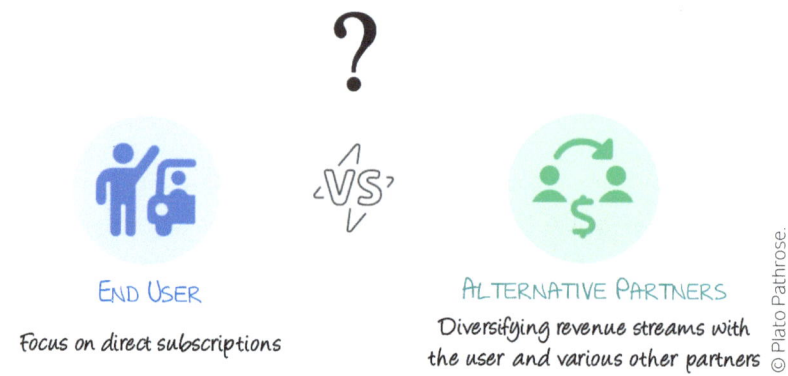

The main lifecycle of the vehicle where the vehicle manufacturer can target in generating revenue is during the operation phase and the service and maintenance phase. Analyzing these lifecycle phases, it can be identified that many stakeholders are there who interact directly or indirectly with the vehicle. These stakeholders can be considered as a revenue source for the vehicle manufacturer if certain partnerships and engagements are arranged.

All these partners or stakeholders may not have a direct influence on the user or the vehicle at all times over different lifecycle phases. A brief overview of various stakeholders and partners and the possible ways of generating a partnership-based revenue is presented in **Figure 9.5**. The advantage of this business model, which diversifies the revenue streams, is that it is not focused only on the vehicle user; Instead, different stakeholders contribute who benefit from the vehicle and the user, which acts as the revenue stream for the vehicle manufacturer.

Figure 9.5 Diverse revenue streams for SDVs.

9.6.
Digitalization and Ecosystem-Driven Innovation

In previous chapters, we discussed the transformation of the mobility industry with digitalization and adaptation of many technology enablers. Most of this transformation was driven by the influence of other industries and evolution of digital technologies. The possibility of building digital environments and prototypes drives many advancements, including the possibility to approach design and development with data-driven decisions based on digital analysis. With various new mobility concepts, such as MaaS, on-demand services, etc., many digital platforms and technology enablers have become the foundational framework in providing various facilities to their users.

The mobility industry is driven mainly by the advancements in the ecosystem rather than the technological evolution within the automotive industry alone. Technologies used successfully in other industries and the concepts of open application and usage are becoming part of the mobility industry now. Connectivity, usage of data platforms, service-based business models, and offering of subscription-based services are adopted from other industries. What caused these changes?

It is driven by the open approach of adopting technologies from other industries and, with systems thinking, applying different concepts and technologies from these industries to the mobility industry. The fundamental need for shared insights and data for the benefit of the product and customers acts as the foundation for change in mentality and different approaches taken by various mobility players these days. The evolution and power of software, together with enablers such as AI and data platforms, drive today's technologies in the mobility industry. All these contribute to a new era of mobility driven by a digital ecosystem and shared insights from different industries to serve the customer. Since digitalization is the foundational framework across multiple industries, integrating various industries and services across them provides an opportunity to adopt new business models and provide services, thereby facilitating a safe and secure ecosystem for the well-being of people.

9.7.
Time to Market—The Ultimate Goal and Challenges

What is the greatest goal for a vehicle manufacturer these days? It is about building a safe and secure vehicle with fantastic functions and delivering it to the customer in a minimum amount of time. Customers are not willing to wait, unlike in the past, when it took five to six years for a new vehicle to be developed and deployed in the market after a vehicle manufacturer announced it. Some vehicle manufacturers now take 10 to 12 months to deploy a vehicle, whereas some take about four to five years. Time to market plays a critical role for all vehicle manufacturers these days. With all these digitalization and advanced technologies that are well supported by enablers like AI and data platforms, a lack of efficiency will result if vehicle manufacturers cannot adapt to the changing ecosystem and they cannot use those resources to improve their vehicle development and deployment timings.

Vehicle manufacturers are addressing this time-to-market challenge by establishing vehicle platforms, building hardware as a separate stream from software that can enrich the vehicle with various functions, and utilizing the latest technologies and enablers across all vehicle development and deployment processes. Strategic partnerships, collaboration, and standardization are crucial in this game. Even though the vehicle manufacturer cannot build everything in a vehicle program, it is necessary to have good partners who are strategically aligned with the goals and targets of the vehicle manufacturer. It should also be a win-win approach for the supplier to work on a specific platform and solutions that could be part of the vehicle platform or part of the vehicle software. Vehicle OS initiatives and open-source projects to build hardware abstraction layers by collaborating are examples of these strategic collaborations observed in the market.

For vehicle manufacturers and their suppliers, there has been a transformation in the project management approach nowadays with SDVs. These days, project management approaches and their deliverables are driven by sales, rather than a predefined set of features and goals that were put together as part of the plan. This change in approach was driven by the dynamic nature that software can bring to vehicles and how businesses are adapting to the quickly changing market. It gives flexibility to decide which features should go with the vehicle during its launch and which should be planned later, according to each geographical region, and based on the demand.

One of the challenges we see in current vehicles is that a hybrid approach is followed for SDVs, and there is a significant challenge of not having enough headroom for vehicle hardware to incorporate changes and provide specific services in the future, for instance three to four years after the vehicle is sold. When focusing on quick product development and deployment with constraints like time and cost of the vehicle, it may not be possible to incorporate high-end hardware that can run for another 20 years, with the possible ingestion of software in the future to undergo upgrades by adding new features and updates. If that happens, the vehicle, which was dynamic with changes and software updates for the first few years, becomes static just after a few years it reaches the user.

The obsolescence of vehicle systems is a complex situation that is challenging for both vehicle manufacturers and users. This is usually addressed by strategically integrating hardware with enough headroom for scalable software deployment for the next five to seven years and also integrating connectivity-based solutions, where the dynamic behavior can still be supported by operating software in the cloud platform-based ecosystem rather than directly deploying it in vehicle systems. Many vehicle manufacturers also address this with a recycling hardware approach, where vehicle systems are replaced after a few years with a new one, adding new features and functions. This involves cost, and it is yet to be identified as a reasonable and sustainable business model. We have to wait and see which business model is profitable after a few years, once vehicle manufacturers' data on these approaches are made available to the public.

9.8. Summary

We are in the world of evolving technologies in the mobility sector. This chapter discussed mobility changes in the future and various business models implemented by different players in the mobility sector. There is a drastic change in people's mindset from owning a vehicle to using transportation as a service, which is driven by the changing ecosystem and living conditions across the world. With all these changes, it is becoming a challenge for many vehicle manufacturers to convey the benefits and value additions of their service or the vehicle to make customers continue with the subscription models. Personalization and the different ways in which it can be applied in the vehicle have also been discussed.

Along with personalization and the changing mobility ecosystem with new SDVs that are influenced by digitalization, different business models are available in the market these days. How various models are deployed in the market, identifying their strengths in building up the business, and the challenges encountered while scaling them up have been discussed in detail. Time to market has become a critical goal in today's mobility industry, and it is not easy for every vehicle manufacturer to address this change in the automotive industry. The various challenges that vehicle manufacturers face these days while moving toward SDVs and solutions to them have been discussed. It is still a very ambiguous environment in the mobility industry to propose which business model will be successful in these highly competitive and changing technology ecosystems. We can be happy that we are in an exciting time with many technologies evolving and getting adopted at lightning speed in the automotive industry. Software is not just a component of the vehicle; instead, software, together with all technology enablers, has penetrated every aspect of our lives and is providing fantastic benefits that we could not have imagined so far. Let us prepare ourselves to witness this software magic that takes us to the future of mobility!!!

References

9.1. Kubitz, B. and Gleave, J., *Mobility as a Service: Its Development, Deployment, and Future* (London, UK: Institution of Engineering and Technology, 2024).

9.2. MaaS-Alliance, "What Is MaaS?," n.d., accessed April 19, 2025, https://maas-alliance.eu/homepage/what-is-maas/.

9.3. MHP Management- und IT-Beratung GmbH, "Enter the Car Subscription Market: Car Subscriptions – The Right Implementation Strategy as a Success Factor in a Scaling Business Model," MHP – A Porsche Company, Ludwigsburg, Germany, September 2021, accessed April 28, 2025, https://www.mhp.com.

Conclusion

The software is not just a tool; it is a force, a limitless engine of imagination, capable of breathing wonder into every corner of the vehicle. Yet, this vision only becomes reality when the software perfectly syncs with the hardware, with every mechanical beat of the vehicle's heart. This story, what you read, may have been told from the perspective of software, but let the truth echo loud and clear.

Software alone cannot carry the dream and cannot exist in isolation. It is the seamless union of code and machine, the flawless integration of every part, that unleashes the true power of the Software-Defined Vehicles.

The Story of Software-Defined Vehicle Continues...

Index

A

Advanced driver assistance systems (ADAS), 5, 110
Agile/DevOps methods, 142, 152
ARM Holdings plc., 104
Artificial Intelligence (AI), 4, 115–116, 120, 128–129, 147, 165–168, 172–173
ASPICE, 147
Automated driving, 60
Automotive ecosystem, 54
Automotive industry, 137, 138, 148–150, 158, 172, 173
Automotive safety integrity levels (ASIL), 128

B

Bandwidth utilization, 150
Bayerische Motoren Werke plant in Dingolfing, 60
Bill of material (BoM) costs, 39
Business models
    and challenges, 182–185
    OEM, 187–189

C

Classical electronic system, 2
Cloud-based data platforms, 168–170
Connected Vehicle Systems Alliance (COVESA), 105
Connectivity layer, 170–171
Consumer perspectives, 5–6
Controller area network (CAN), 82
CPSs with ML algorithms, 47
Cyberattacks, 50
Cyber-physical system (CPS), 10, 14, 18–22, 99–100
Cybersecurity
    AI components, 120
    best practices, 121
    connectivity options, 117
    control and secure attack surfaces, 118
    controlling remote access and configuration, 119
    definition, 117–118
    investing and controlling organizational security processes, 120
    regulation and compliance, 121
    SDCAVs
        analysis of, 122–124
        benefits, 126
        ISO/PAS 8800 Road Vehicles, 129
        ISO/PAS 21448 Road Vehicles, 128
        ISO 21434 Road Vehicles, 129
        ISO 26262 Road Vehicles, 128
        operational safety, 129–130
        product safety, 130
        risks and vulnerabilities, 125
        two-zone analysis model, 125, 126
    software supply chain, 118–119
    software updates and upgrades, 119–120
Cybersecurity risks, 74

D

Data center, 161–164
Data-driven design approach, 160
Data monetization, 185–186
Data platform layer, 83
Data storage system for autonomous driving (DSSAD), 83
DevOps, 57
Digital engineering, 139–140, 143, 149, 150
Digital-first shift, 158–161
Digital twins, 137, 138, 146, 149
Digitalization, 136–140, 190, 191
Domain-centralized architecture, 85, 86
Domain controllers, 16, 17
Downloaded packages, 42

E

Eclipse Foundation, 104
Eco-friendly transportation, 6
Ecosystem and technological evolution, 5–6
Ecosystem-driven innovation, 190
Electrical and electronics (E/E) architecture, 82
    distributed architecture, 85, 86
    domain-centralized architecture, 85, 86
    hybrid architecture, 87–89
    platform approach, 89–91
    product development, 162–163
    zonal architecture, 85–87
Electronic control units (ECUs), 16, 17, 86, 87
Enablers
    Artificial Intelligence, 165–168
    cloud-based data platforms, 168–170
    connectivity, 170–171
    software and digitalization concepts, 55
Enabling systems (ESs), 23, 24
End of Line (EoL) calibration, 153
Event data recorder (EDR), 83

E

Exciting needs/delighters, 6–9
External sensors, 61

F

Feature-on-demand (FoD), 41, 119–120
Firmware over-the-air (FOTA), 41
Function-driven development (FDD), 142

G

General-purpose OS (GPOS), 99
Generational change, 179
Generative AI (GenAI), 139, 147, 165–166, 173
Gigabit Ethernet (GigE), 87

H

Hardware Abstraction Layer for Software-Defined Vehicles (HAL4SDV), 105–106
Hardware-driven automotive industry, 159
Hardware/mechanical systems, 47
Hazard and risk analysis (HARA), 112

I

Information technology perspective, 51
Infrastructure and connectivity dependencies, 75
Innovation, 10
Intellectual properties (IPs), 119
Internal Combustion Vehicle (ICE), 171
ISO/IEC/IEEE 15288 systems and software engineering, 14
ISO/PAS 8800 Road Vehicles, 129
ISO/PAS 21448 Road Vehicles, 128
ISO 21434 Road Vehicles, 129
ISO 26262 Road Vehicles, 128

K

Kano model, 6, 7
Kappa architecture, 97–98

L

Lambda architecture, 97–98
Large Language Models (LLMs), 165–166
Legacy manufacturers, 87
Local interconnect network (LIN), 82

M

Machine language (ML), 115–116
Manufacturer and vehicle lifecycle phases, 54–56
Micro-mobility solutions, 5
Microservices architecture, 95–96
Mobility as a service (MaaS), 179–180, 183
Mobility ecosystem, 178
Mobility solutions, 27
Model-based development, 145–146
Modular-monolithic architecture, 92–93
Monetization and return on investments (RoI), 76
Monolithic architecture, 91–92
Multimodal and on-demand mobility ecosystem, 179–180
Multimodal transportation, 5, 180

N

Natural language processing (NLP), 165
Navigation maps, 37
Normal needs, 8

O

Obsolescence of vehicle systems, 192
On-demand storage, 146
Open-source software, 102–103
Operating systems (OSs), 20, 98–101
Over-the-air (OTA) updates, 37, 39–43, 114–115, 120–121, 169
    architecture, 41, 42

P

Personalization features during journey, 181, 182

Predictive maintenance, 46, 51
Predictive personalization, 182
Product deployment, 134–136
Product development, 134–140
    development platform and data center, 161–164
    E/E architecture, 162–163
    hardware-first to digital-first
        data-driven design, 160
        decision-making, 161
        digital ecosystem, 160
        performance, quality, and changing UX, 158
        powershift graph, 158, 159
        software development and deployment, 159
        tools, 159–160
    overview, 157
    testing
        Artificial Intelligence, 165–168
        changing skill requirements, 171–173
        cloud-based data platforms, 168–170
        connectivity, 170–171
        technology adoption, 171–173
Product safety, 130

R

Real-time OS (RTOS), 99
Recycling hardware approach, 192
Regulatory and compliance challenges, 75
Regulatory frameworks, ecosystem, 65–67
Remote debugging, 60
Remote diagnostics, 46–47, 60
Research and software development, 11

S

Safety
    challenges for, 112–114
        AI components, 115–116
        OTA updates, 114–115
        security vulnerabilities, 114
        software complexity, 115

supply chain and supplier ecosystem, 116
user, 116–117
during the journey, 178
overview, 110–112
SDCAVs
  analysis of, 122–124
  benefits, 126
  ISO/PAS 8800 Road Vehicles, 129
  ISO/PAS 21448 Road Vehicles, 128
  ISO 21434 Road Vehicles, 129
  ISO 26262 Road Vehicles, 128
  operational safety, 129–130
  product safety, 130
  risks and vulnerabilities, 125
  two-zone analysis model, 125, 126
Safety of the Intended Functionality (SOTIF), 110
Scalable Open Architecture for Embedded Edge (SOAFEE), 104–105
SDCAVs. See Software defined connected and autonomous vehicles (SDCAVs)
Service and maintenance phase of vehicle, 61–64
Service-oriented architecture (SOA), 94–95
Shift-left and shift-right approaches, 151
  agile/DevOps methods, 142, 152
  AI-based tools, 152, 154
  automotive industry, 148–150
  decision, 144, 145
  development ecosystem, 141
  digital engineering, 143
  digitalization, 136–140
  DMAIC methodology, 144
  EoL calibration, 153
  lifecycle phase, 152
  macrolevel approach, 143
  product development and deployment, 134–136
  production phase, 144
  SIPOC structure, 140, 141
  software testing, 142

tools and techniques, 145–147
vehicle manufacturing process, 153
Smartphone *versus* today's vehicles, 2, 3
SOAFEE Special Interest Group (SIG), 104
Software
  changes to vehicles and service, 12
  complexity, 75
  definition of, 3–4
  functions of, 32–34
  quality and rollback possibilities, 37
  reliability, 75
Software defined connected and autonomous vehicles (SDCAVs), 22
  advantages and benefits, 50
  bottom-up decomposition, 81
  challenges and restrictions, 74–76
  COVESA, 105
  CPSs, 82, 83
  data platform layer, 83
  development, 80
  Eclipse SDV working group, 104
  E/E architecture layer, 82
    distributed architecture, 85, 86
    domain-centralized architecture, 85, 86
    hybrid architecture, 87–89
    platform approach, 89–91
    zonal architecture, 85–87
  engineering and business aspects, 28
  ethernet, CAN, and LIN, 82
  five-layer logical architecture, 80, 81
  HAL4SDV, 105–106
  lambda and kappa architectures, 97–98
  lifecycle phases, 71
  microservices architecture, 95–96
  mobility in various industries, 28
  modular-monolithic architecture, 92–93

monolithic architecture, 91–92
open-source software, 102–103
operating system, 98–101
profitable models over time, 28
safety and cybersecurity
  analysis of, 122–124
  benefits, 126
  ISO/PAS 8800 Road Vehicles, 129
  ISO/PAS 21448 Road Vehicles, 128
  ISO 21434 Road Vehicles, 129
  ISO 26262 Road Vehicles, 128
  operational safety, 129–130
  product safety, 130
  risks and vulnerabilities, 125
  two-zone analysis model, 125, 126
SOA, 94–95
SOAFEE, 104–105
software platform layer, 82–83
use cases, 71–73
vehicle connectivity layer, 83
Software defined systems for users, 34–36
Software-defined vehicles (SDVs), 13
  business models and possibilities, 28
  concepts and digitalization, 55
  definition, 23
  efficiency, 67–69
  interdependency of hardware and software, 39
  project management approach, 191
  software influence, 26
  vehicle costs, 69–71
Software-driven mentality, 48
Software-intensive systems (SIS), 18–22
Software over-the-air (SOTA), 41
Software platform layer, 82–83

Software updates, 17, 36
  debugging and fixing faults, 36
  dependency on connectivity and storage of private data, 50
  for each electronic component in vehicle, 25
  over-the-air (OTA) updates, 37, 39
  software deployment processes, 49
  strategy, 38, 49
Supply chain and supplier ecosystem, 116
Supply chain dependencies, 76
Sustainable transportation, 6
Systems engineering, 14
  domains and functionalities, 16

E/E architecture, 18
  enabling systems, 23
  functionalities and interfaces, 26
  lifecycle phases, 14, 15
  micro- and mini-computers, 16
  vehicle transformation, 15

T
Test-driven development (TDD), 142
Time-to-market challenge, 191–192

U
United Nations Economic Commission for Europe (UNECE) regulations R155 and R156, 40

Urbanization and global population, 179

V
Vehicle connectivity layer, 83
Vehicle design strategy, 40
Vehicle development and deployment timings, 191
Vehicle's operation phase, 45
V-model, 112, 134, 135, 142

W
Wireless connectivity, 27
"WOW" factor from vehicle, 43–45, 51

Z
Zonal controllers, 16

About the Author

Plato Pathrose

Plato Pathrose holds a degree in Electronics and Communication Engineering from the University of Kerala, India. He began his career as a Design and Development Engineer, contributing to safety-critical systems in the automotive and avionics sectors. Over the years, he has led numerous systems engineering teams, overseeing projects across various phases of the product lifecycle. His professional journey spans two decades with multiple levels of the automotive industry, from original equipment manufacturers (OEMs) to Tier 1 and Tier 2 suppliers, providing him with deep insights into emerging technologies, development processes, and evolving business models. As a consultant, he has worked with a diverse range of organizations, including government bodies, supporting the design and development of safety-critical systems and shaping concepts and requirements for software-defined vehicles (SDVs) and their integration with infrastructure and public deployment.

Plato is the founder of PTS Creative, a consulting and technology services firm based in Germany, and a techno-strategic advisor to organizations in the new mobility area, particularly in advanced driver assistance systems (ADAS), automated driving, digitalization, and SDVs. Beyond his technical expertise, he is a specialist in processes and methodologies, a certified Lean Six Sigma expert, and a practitioner of TRIZ innovation methodologies. He offers coaching and consulting services to help organizations optimize their product development and production processes. He is a certified Automotive SPICE Assessor, Project Management Professional, and Agile Coach who helps organizations transform their operations to integrate agile methodologies with digitalization. He is a technology speaker with technical publications and various technology books to his name, laying the foundation for the development and testing of autonomous vehicles worldwide. He was born and brought up in Trivandrum, a city in the southern state of Kerala in India, and lives in Karlsruhe, Germany, along with his wife and son. He is a passionate musician, a professional flutist, and an avid traveler.

Email: platopathrose@gmail.com
Website: www.platopathrose.com